AF248589

THE **HISTORY** *of* **MEDICINE**

MEDICINE BECOMES A SCIENCE

1840–1999

THE HISTORY *of* MEDICINE

MEDICINE BECOMES A SCIENCE

1840–1999

KATE KELLY

Facts On File
An imprint of Infobase Publishing

MEDICINE BECOMES A SCIENCE: 1840–1999

Copyright © 2010 by Kate Kelly

Facts On File, Inc.
An imprint of Infobase Publishing
132 West 31st Street
New York NY 10001

Library of Congress Cataloging-in-Publication Data

Kelly, Kate, 1950–
    Medicine becomes a science : 1840–1999 / Kate Kelly.
        p. ; cm.—(History of medicine)
    Includes bibliographical references and index.
    ISBN 978-0-8160-7209-5 (alk. paper)
    1. Medicine—History—19th century. 2. Medicine—History—20th century.
I. Title. II. Series: History of medicine (Facts on File, Inc.)
    [DNLM: 1. History of Medicine. 2. History, 19th Century. 3. History, 20th
Century. 4. Science—history. WZ 40 K29m 2010]
    R149.K45 2010
    610.9—dc22   2009011598

Facts On File books are available at special discounts when purchased in bulk quantities for businesses, associations, institutions, or sales promotions. Please call our Special Sales Department in New York at (212) 967-8800 or (800) 322-8755.

You can find Facts On File on the World Wide Web at http://www.factsonfile.com

Text design by Annie O'Donnell
Illustrations by Bobbi McCutcheon
Photo research by Elizabeth H. Oakes
Composition by Hermitage Publishing Services
Cover printed by Yurchak Printing, Landisville, Pa.
Book printed and bound by Yurchak Printing, Landisville, Pa.

Printed in the United States of America

This book is printed on acid-free paper.

CONTENTS

"You have to know the past to understand the present."
—*American scientist Carl Sagan (1934–96)*

The history of medicine offers a fascinating lens through which to view humankind. Maintaining good health, overcoming disease, and caring for wounds and broken bones was as important to primitive people as it is to us today, and every civilization participated in efforts to keep its population healthy. As scientists continue to study the past, they are finding more and more information about how early civilizations coped with health problems, and they are gaining greater understanding of how health practitioners in earlier times made their discoveries. This information contributes to our understanding today of the science of medicine and healing.

In many ways, medicine is a very young science. Until the mid-19th century, no one knew of the existence of germs, so as a result, any solutions that healers might have tried could not address the root cause of many illnesses. Yet for several thousand years, medicine has been practiced, often quite successfully. While progress in any field is never linear (very early, nothing was written down; later, it may have been written down, but there was little intra-community communication), readers will see that some civilizations made great advances in certain health-related areas only to see the knowledge forgotten or ignored after the civilization faded. Two early examples of this are Hippocrates' patient-centered healing philosophy and the amazing contributions of the Romans to public health through water-delivery and waste-removal systems. This knowledge was lost and had to be regained later.

The six volumes in the History of Medicine set are written to stand alone, but combined, the set presents the entire sweep of the history of medicine. It is written to put into perspective for high school students and the general public how and when various medical discoveries were made and how that information affected health care of the time period. The set starts with primitive humans and concludes with a final volume that presents readers with the very vital information they will need as they must answer society's questions of the future about everything from understanding one's personal risk of certain diseases to the ethics of organ transplants and the increasingly complex questions about preservation of life.

Each volume is interdisciplinary, blending discussions of the history, biology, chemistry, medicine and economic issues and public policy that are associated with each topic. *Early Civilizations,* the first volume, presents new research about very old cultures because modern technology has yielded new information on the study of ancient civilizations. The healing practices of primitive humans and of the ancient civilizations in India and China are outlined, and this volume describes the many contributions of the Greeks and Romans, including Hippocrates' patient-centric approach to illness and how the Romans improved public health.

The Middle Ages addresses the religious influence on the practice of medicine and the eventual growth of universities that provided a medical education. During the Middle Ages, sanitation became a major issue, and necessity eventually drove improvements to public health. Women also made contributions to the medical field during this time. *The Middle Ages* describes the manner in which medieval society coped with the Black Death (bubonic plague) and leprosy, as illustrative of the medical thinking of this era. The volume concludes with information on the golden age of Islamic medicine, during which considerable medical progress was made.

The Scientific Revolution and Medicine describes how disease flourished because of an increase in population, and the book describes the numerous discoveries that were an important aspect of this time. The volume explains the progress made by Andreas Vesalius (1514–64) who transformed Western concepts of the structure of the human body; William Harvey (1578–1657), who studied and wrote about the circulation of the human blood; and Ambroise Paré (1510–90), who was a leader in surgery. Syphilis was a major scourge of this time, and the way that society coped with what seemed to be a new illness is explained. Not all beliefs of this time were progressive, and the occult sciences of astrology and alchemy were an important influence in medicine, despite scientific advances.

Old World and New describes what was happening in the colonies as America was being settled and examines the illnesses that beset them and the way in which they were treated. However, before leaving the Old World, there are several important figures who will be introduced: Thomas Sydenham (1624–89) who was known as the English Hippocrates, Herman Boerhaave (1668–1738) who revitalized the teaching of clinical medicine, and Johann Peter Frank (1745–1821) who was an early proponent of the public health movement.

Medicine Becomes a Science begins during the era in which scientists discovered that bacteria was the cause of illness. Until 150 years ago, scientists had no idea why people became ill. This volume describes the evolution of "germ theory" and describes advances that followed quickly after bacteria was identified, including vaccinations, antibiotics, and an understanding of the importance of cleanliness. Evidence-based medicine is introduced as are medical discoveries from the battlefield.

Medicine Today examines the current state of medicine and reflects how DNA, genetic testing, nanotechnology, and stem cell research all hold the promise of enormous developments within the course of the next few years. It provides a framework for teachers and students to understand better the news stories that are sure to be written on these various topics: What are stem cells, and why is investigating them so important to scientists? And what is nanotechnology? Should genetic testing be permitted? Each of the issues discussed are placed in context of the ethical issues surrounding it.

Each volume within the History of Medicine set includes an index, a chronology of notable events, a glossary of significant terms and concepts, a helpful list of Internet resources, and an array of historical and current print sources for further research. Photographs, tables, and line art accompany the text.

I am a science and medical writer with the good fortune to be assigned this set. For a number of years I have written books in collaboration with physicians who wanted to share their medical knowledge with laypeople, and this has provided an excellent background in understanding the science and medicine of good health. In addition, I am a frequent guest at middle and high schools and at public libraries addressing audiences on the history of U.S. presidential election days, and this regular experience with students keeps me fresh when it comes to understanding how best to convey information to these audiences.

What is happening in the world of medicine and health technology today may affect the career choices of many, and it will affect the health care of all, so the topics are of vital importance. In addition, the public health policies under consideration (what medicines to develop, whether to permit stem cell research, what health records to put online, and how and when to use what types of technology, etc.) will have a big impact on all people in the future. These subjects are in the news daily, and students who can turn to authoritative science volumes on the topic will be better prepared to understand the story behind the news.

ACKNOWLEDGMENTS

This book, as well as the others in the set, was made possible because of the guidance, inspiration, and advice offered by many generous individuals who have helped me better understand science and medicine and their histories. I would like to express my heartfelt appreciation to Frank Darmstadt, whose vision and enthusiastic encouragement, patience, and support helped shape the set and saw it through to completion. Thank you, too, to the Facts On File staff members who worked on it.

The line art and the photographs for the entire set were provided by two very helpful professionals—Bobbi McCutcheon provided all the line art; she frequently reached out to me from her office in Juneau, Alaska, to offer very welcome advice and support as we worked through the complexities of the renderings. A very warm thank you to Elizabeth Oakes for finding a wealth of wonderful photographs that helped bring the information to life. Carol Sailors got me off to a great start, and Carole Johnson kept me sane by providing able help on the back matter of all the books. My agent Bob Diforio has remained steadfast in his shepherding of the work.

I also want to acknowledge the wonderful archive collections that have provided information for the book. Without places such as the Sophia Smith Collection at the Smith College Library, first-hand accounts of the Civil War battlefield treatment or reports such as Lillian Gilbreth's on helping the disabled after World War I would be lost to history.

"If it is a terrifying thought that life is at the mercy of the multiplication of these minute bodies, it is a consoling hope that Science will not always remain powerless before such enemies . . ."

—Louis Pasteur in a paper read before the French Academy of Sciences, April 29, 1878

Only 150 years ago, scientists did not know what made people sick. There were many theories of how and why illness spread, but none of them were accurate. Though very primitive microscopes had permitted the examination of bacteria as early as the 1660s, it was not until the mid-19th century that bacteria's contribution to the spread of illness was understood. *Medicine Becomes a Science* describes the historic events, scientific principles, and technical breakthroughs that led to a century and a half of rapid advancement in combating disease.

Medicine Becomes a Science: 1840–1999 briefly introduces Antoni van Leeuwenhoek, a Dutch cloth merchant who was first to see and identify various forms of bacteria. However, he—and no one else—fully understood what he was seeing. Then in the mid-1800s, Louis Pasteur, a professor of chemistry at Strasbourg University, came up with the concept of germ theory, which was to change the world of medicine forever. The German microbiologist Robert Koch built on this theory by adding his own three laws in 1883. These laws provided a system that led to an understanding of how to identify the organisms that cause disease.

From this time forward, medical progress has moved swiftly. Louis Pasteur himself went on to make other important discoveries. His work on ways to prevent the transmission of rabies was instrumental in laying the groundwork for vaccines—a method of disease prevention we rely upon today.

Students who have grown up being constantly told to "go wash up" will be quite surprised to read about 19th-century surgeons who routinely examined patients in the mornings and then performed surgeries in the afternoon; they wore no gloves and no one thought of hand washing between activities. The physician Ignaz Semmelweis made the connection between the lack of cleanliness and the spread of infection, but few changes were made until the Scottish physician Joseph Lister came along and pushed for greater sanitation in hospitals.

In the 1890s, scientists came to suspect the existence of viruses as causative agents for some diseases. They had to accept this "discovery" on faith as they did not yet know about viruses because technology powerful enough to view them was not created until the 1930s.

Women have always played a critical role in health care, but they generally worked behind the scenes, caring for family members at home and helping with the births of friends and family. In the 19th century, women began making major contributions to medicine. The field of nursing was established as a professional field, and women began breaking barriers to become both physicians and scientists. This paved the way for women of today who are active in all areas of medical science.

Medicine Becomes a Science: 1840–1999 helps readers understand the medicine of today. In the 160 years covered by this volume, medical knowledge surged forward, and the information is illuminating. The back matter contains a chronology, a glossary, and an array of historical and current sources for further research. These sections should prove especially helpful for readers who need additional information on specific terms, topics, and developments in medical science.

Independent thinking is often an important part of scientific inquiry, and this is well illustrated by the story of the bacteriologist Alexander Fleming's discovery that mold could be grown and used to fight deadly illnesses. Jonas Salk's work to eradicate polio further illustrates how a brilliant mind can solve a problem.

Today, medical diagnosis and treatment follow what has come to be called evidence-based medicine, which involves integrating individual clinical expertise with the best available evidence from systematic research. Scientists and medical practitioners rely on science, engineering, the statistics from studies and randomized-control trials, before they choose the medical treatment that seems best for each individual.

Readers of this volume will come away with an understanding of the state of medical care as it existed before the 21st century. Chapter 1 describes the stunning discoveries made by Louis Pasteur and Robert Koch that finally provided an understanding of what caused disease. Germ theory opened a whole new world in medicine by creating a way for physicians to do more than offer palliative care. Chapter 2 introduces women's contributions to medicine, including information about the first woman doctor, the founding of the profession of nursing, and the contributions of women like Florence Nightingale and Clara Barton. Chapter 3 describes how X-rays were discovered and notes the contributions of Marie and Pierre Curie. Chapter 4 highlights the accidental discovery of penicillin, a medicine that became a vital part of doctors' weapons against disease. Chapter 5 focuses on polio and explains how Jonas Salk and Albert Sabin both contributed to the eradication of the disease. The chapter continues with a description of the new ideas behind evidence-based medicine—ideas that have resulted in a new and more scientific way of looking at disease. Chapter 6 examines the aftermath of 20th-century warfare and what it meant for the disabled. For the first time, considerable numbers of soldiers were surviving major injuries, and this provided the impetus for improving treatment of people who returned from war but had to cope with some type of handicap. Chapter 7 traces how scientists came to understand the science of the blood and continues with information about artificial hearts, heart transplants, and what is known about heart disease. Chapter 8 looks at medicine in the late 1990s, how diagnoses and treatments have been influenced by the discovery of DNA.

This book is a vital addition to literature on the history of medicine because it puts into perspective the medical discoveries of the period and provides readers with a better understanding of the accomplishments of the time. During this period, scientists and physicians finally realized the cause of disease, and, with this discovery, medical progress began flying forward.

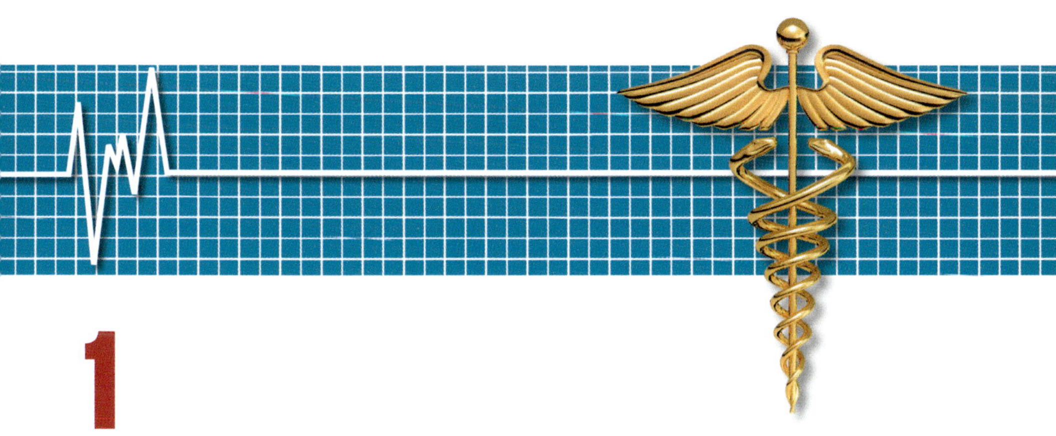

1

Medical Science Finally Advances

Well into the 1800s, physicians continued to believe that miasmas, gases that arose from sewers, swamps, garbage pits, or open graves (and were thought to be poisonous), caused illness. Because physicians did not yet understand the cause of illness, guesswork was heavily involved when treatments were devised. Doctors frequently used leeches to bleed patients, and laxatives, opium, peppermint, and brandy were often considered cures. While some herbal medications have come back into favor today, many of the medicines used early on—mercury among them—are now known to be poisonous or cause serious, if not fatal, damage.

Many important steps preceded the identification of bacteria and its role in causing diseases. The invention of the microscope was key, and, while microscopes were actually invented by scientists who were working late in the 1500s, it was a hobbyist who was actually the first to see bacteria. Antoni van Leeuwenhoek (1632–1723) was a Dutch cloth merchant who ground his own magnifying lenses to more carefully inspect the fabric he was purchasing for his business. He succeeded in creating more powerful lenses than the microscopes created by Robert Hooke in England

and Jan Swammerdam in the Netherlands; they created magnification that enlarged objects only 20 to 30 times. Leeuwenhoek's grinding skill, his acute eyesight, and his intuitive understanding of how to direct light onto the object permitted him to see items that were estimated to be 200 times their natural size.

Leeuwenhoek was fascinated by the world he could see through his lenses, and as a hobby he studied animal and plant tissues as well as mineral crystals and fossils. He was the first to see microscopic animals such as nematodes (roundworms) and rotifers (multicelled animals that have a disk at one end with circles of strong cilia that often look like spinning wheels), as well as blood cells and living sperm. Leeuwenhoek created careful descriptions of exactly what he saw, and he hired an illustrator to draw what he described. He began submitting his information to the Royal Society of London, an organization founded in 1660 to bring attention to science scholarship. Because of his meticulous efforts and his unique discoveries, the Society recognized the merit of this cloth merchant's work and began to publish it. Leeuwenhoek's articles often took precedence over work from credentialed scientists. They had his descriptions translated from Dutch into English or Latin, and his findings were regularly published in the Society's publication.

The next person with a theory that might have moved the science of germ theory forward was totally ignored. In the 16th century, Giralamo Fracastoro (1478–1553) came up with the concept that diseases were caused by living organisms that were too small to see, but Fracastoro's ideas died with him. Finally, in the early 19th century, progress began to be made. Rudolf Virchow (1821–1902) determined that cells were not only the basic unit of life, but also the basic unit for disease. A few years later, Louis Pasteur (1822–95) developed a way to more fully explore and verify germ theory. From Pasteur's work, microbiologist Robert Koch (1843–1910) formulated the rules that helped scientists evaluate the cause of disease.

This chapter sets the scene for how and why medical progress began to happen quickly in so many areas. The work of Pasteur and Koch finally explained the cause of disease, but it required many others to help bring about clinical change. Ignaz Semmelweis (1818–65) was the first to realize that the unclean atmosphere during surgery was the likely cause of many infections, but his abrasive personality hindered his ability to encourage others. Joseph Lister (1827–1912) was the right person to push for a cleaner environment for surgery.

LOUIS PASTEUR (1822–1895): MAJOR ADVANCES IN MEDICINE

Louis Pasteur contributed in multiple ways to the advancement of science. He began to realize that disease was communicable and that illnesses were spread by tiny microorganisms called germs. Ultimately, Pasteur went on to demonstrate that microscopic organisms could cause illnesses.

To understand Pasteur's contributions, it is important to recall that spontaneous generation was still considered a viable explanation for the presence of any entity that appeared where it had not been previously. This theory was countered somewhat by work done in 1668 by the Italian biologist Francesco Redi (1626–97), who successfully challenged spontaneous generation with an experiment involving maggots and meat. When he covered the meat with gauze to prevent flies from laying their eggs on it, no maggots appeared on the meat. (The maggots they talked of were actually fly larvae, which hatch from flies' eggs.)

By the 19th century, spontaneous generation was hotly debated. While scientists were beginning to believe that maggots, mice, and worms could not generate spontaneously, they still had no other explanation for the microscopic animals that were now visible through microscopes. The topic was very much under discussion, so the Paris Academy of Sciences offered a prize for any experiments that would help resolve the conflict.

Debunking Spontaneous Generation

From his work, Pasteur had come to believe that decay in organic matter was caused by germs—now recognized as *microbes*—that

Louis Pasteur was one of the greatest scientists of all times. *(Dibner Library of the History of Science and Technology, Smithsonian Institution Libraries)*

floated in the air. Pasteur devised a method to study whether microbes could generate spontaneously. He based his experiment on a number of previous accepted observations. Scientists knew bacteria grew in open containers of meat broth, and they accepted that bacteria would not grow in broth in an airtight container. Pasteur reasoned that if bacteria could generate spontaneously, then something that provided the right environment (the broth) with something that permitted air to enter would demonstrate whether spontaneous generation was possible.

Pasteur wanted a container that would allow air to flow in and circulate but would keep other matter from entering, so he selected a glass flask with a long thin, curved neck for his experiment. He carefully sterilized the flask and poured a nutrient broth (a clear soup) into it. Next he boiled the broth to kill any living matter that might have started out in the liquid. The sterile broth was then left to sit at room temperature. After several weeks, Pasteur noted that the broth in the curved-neck flask exhibited no change at all—bacteria, which everyone agreed were in the air, had not spontaneously generated when the air entered the flask. Though air flowed in freely, germs were trapped in the curved neck of the flask, and this prevented them from reaching the broth. If germs could generate spontaneously, then of course they would have grown in the broth.

In 1864, Louis Pasteur received the prize for devising an experiment that definitively proved that microorganisms are present in air, but that air cannot give rise to organisms spontaneously, finally putting the argument about spontaneous generation fully to rest. Pasteur, who had previously been rejected by the Academy of Sciences, was now admitted.

Using Science to Address Practical Problems

Pasteur was well regarded by people in science, business, and government. When the wine industry, extremely important to France's economy, ran into fermentation problems, Emperor Napoléon I Bonaparte (1769–1821) personally stepped in to ask Pasteur to apply his scientific knowledge to help winemakers. Pasteur worked with heating the wine just enough to kill most of the microbes present and found that chilling the wine kept any remaining microbes from multiplying. Pasteur later learned that this process, which is now called *pasteurization,* could also prevent milk from turning sour and could be used with other food as well.

The next industry to seek Pasteur's help was the silk industry. Output was down because of a disease that was affecting the eggs of the silkworm and reducing their numbers. In 1865, Pasteur identified a microscopic parasite that was infesting the silkworms and the leaves they fed on and showed that by destroying the infected ones the silk industry could be saved. He also devised a method that farmers could use to tell where infection resided so that diseased silkworm eggs could be eliminated from their nurseries.

PASTEUR AND THE MICROBIOLOGIST ROBERT KOCH WORK ON ANTHRAX

The idea of applied science—science used to help overcome problems—became fashionable, after Pasteur proved successful at helping with several industry problems in France. One of the next problems brought to the attention of scientists occurred in Ger-

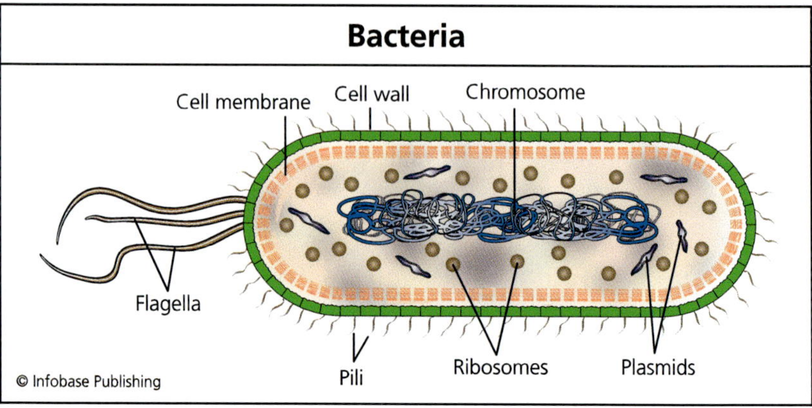

Although Antoni van Leeuwenhoek had seen and described bacteria in the 17th century, it was not until the 19th century with the work done by Louis Pasteur and Robert Koch that there was any conclusive understanding that bacteria were the root cause of many diseases.

many, and the scientist who was approached was a young microbiologist and physician named Robert Koch (1843–1910). In the early 1870s, farmers in Germany were having a terrible problem with anthrax, a devastating disease that was killing their cattle. (The term *anthrax* comes from the Greek word *anthrakitis,* meaning "anthracite," which is coal, in reference to the black skin lesions that develop with some forms of the illness.)

In the late 19th century, anthrax was a major problem. It caused a deadly and highly communicable disease in animals. The spore that caused the disease was hardy and could live a long time. An entire herd of cattle could be infected by walking over the ground where an infected animal had died. The only hope of preventing the spread of the disease was to kill any infected animals and bury them deep in the ground, something that was not easy to do in the winter. (See the sidebar "Anthrax: Modern Weapon in Bioterrorism" on page 11 for information on how terrorists are trying to benefit from the hardiness of the spores.)

Robert Koch was aware of Pasteur's ideas about germs and the work Pasteur had done in the wine and silk industries, and Koch was interested in helping the farmers. He set up a laboratory in his home and began investigating blood samples from the affected cattle. Through the microscope, he identified rod-shaped bacilli as the sign of anthrax, and he began to track the anthrax life cycle by infecting mice with the disease and studying the changes in the infected blood. (Robert Koch's work was the first proof that diseases could be caused by microbes.)

An Anthrax Vaccine

A vaccine had been created to prevent smallpox, and this seemed a logical course of action with the anthrax. However, Edward Jenner (1749–1823) had been able to use the weaker cowpox to inject humans in order to create the antibodies to fight against the more

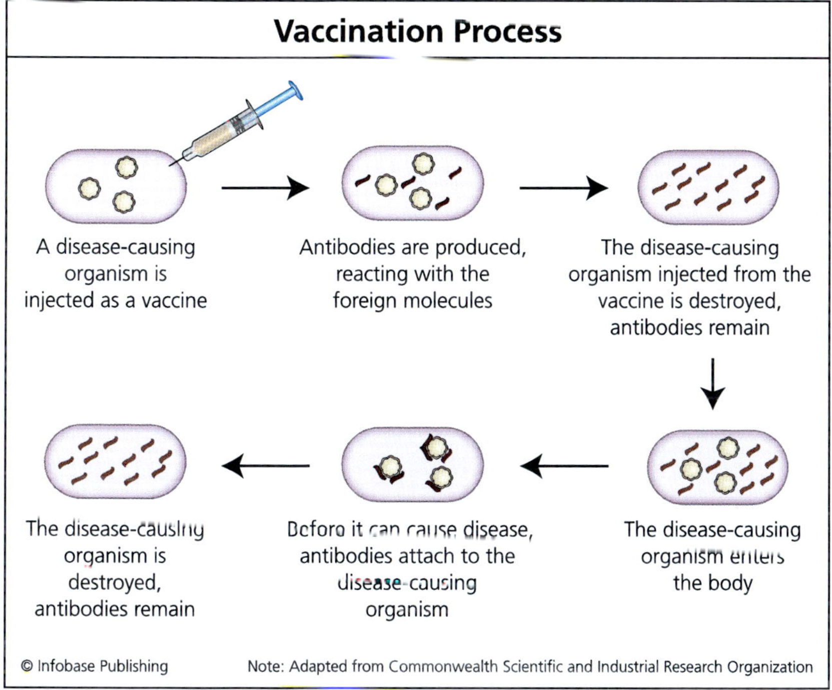

© Infobase Publishing   Note: Adapted from Commonwealth Scientific and Industrial Research Organization

The process of vaccination was first used with smallpox and, although scientists were now finding ways to vaccinate against some other diseases, they did not yet have a scientific understanding of why vaccines worked.

deadly smallpox. Scientists knew of no weaker form of anthrax, and injecting anthrax itself, even in small amounts, would have been deadly. Robert Koch's documentation of the anthrax life cycle and his determination that disease is caused by microbes were very important, but the livestock vaccine to prevent against anthrax resulted from work done by Louis Pasteur. Pasteur's continued experimentation had revealed that with some illnesses, a weakened form of the microbe could be used to immunize against more virulent forms.

Pasteur decided that heat might weaken the germs, so he heated some of the anthrax to reduce virulence and then vaccinated the sheep. He also anticipated the need for a control group. He prepared two groups of 25 sheep, one goat, and several cows. The animals of one group were injected with an anti-anthrax vaccine prepared by Pasteur twice at an interval of 15 days; the control group was left unvaccinated. Thirty days after the first injection, both groups were injected with a culture of live anthrax bacteria. All the animals in the nonvaccinated group died, whereas all in the vaccinated group survived. Though Pasteur perfected this vaccine in 1881, a vaccine for humans was not created until 1954.

PASTEUR EXTENDS WORK ON IMMUNIZATION

Pasteur continued to explore immunization. Rabies was a terrible problem at that time and was fatal to both animals and people. He began his rabies experiments using animals. The study of rabies was time-consuming as there was generally a long delay of several weeks between the time an animal was bitten and the germ reached the brain. However, Pasteur began to develop the idea that the longer reaction time might mean that vaccination could be given within a certain time period after the bite. The other benefit here was that only those who had been bitten by a rabid animal needed to be treated. Pasteur had not yet used the vaccine on humans, but in 1885 a small boy who been bitten by a rabid dog was brought to Pasteur's laboratory. He knew the boy would die if nothing were done, so he administered the vaccine. Several tense weeks later, he knew the vaccine had worked.

In 1888, the Pasteur Institute was founded in France as a clinic for rabies treatment, a research center for disease, and a teaching institute. When a Pasteur Institute was founded in Saigon in 1891, it became the first in a world network and showed the esteem in which his work was held.

KOCH'S POSTULATES

In addition to investigating anthrax, Koch continued studies of various other types of diseases. He made notable inroads into creating a theory of contagion, and in 1883 he set out three laws that explained the cause of disease. Koch's *postulates* have been used ever since to determine whether an organism causes a disease and are as follows:

Robert Koch's postulates created a framework for assessing each disease that was studied.

1. The suspected germ must be consistently associated with the disease.
2. It must be isolated from the sick person and cultured in the laboratory.
3. Experimental inoculation with the organism must cause the symptoms of the disease to appear.

In 1905, a fourth rule was added:

4. Organisms must be isolated again from the experimental infection.

(continues on page 12)

ANTHRAX:
Modern Weapon in Bioterrorism

As the farmers who consulted Robert Koch came to realize, one of the challenges of anthrax is that it can form long-lived spores that are capable of surviving in a hostile environment. The bacteria become *dormant* but can remain viable for decades and perhaps centuries. When anthrax-infected animal burial sites have been disturbed as many as 70 years after the fact, spores have been known to reinfect living animals. (Today, anthrax infections in domestic animals are relatively rare because of animal vaccination programs and sterilization of waste materials. While the disease is most common in animals, it can be transferred to humans. Some forms are so dangerous that a person who has been exposed needs to be quarantined.)

Exposure used to be primarily by occupational exposure to infected animals or their products (usually wool or meat)—the more dangerous form of anthrax used to be called wool sorters' disease. The exposure to this version is via inhalation, and it is very rare. In 2006, a musician who had brought African goatskins to make drums into the United States became very ill from exposure to the anthrax spores on the unprocessed skins. Hospitalized for a month, the 44-year-old victim was able to return to performing within a few months. Prior to this time, the last known case in the United States was in California in 1976 when a home weaver died after working with wool imported from Pakistan. The spores are so deadly that it was very dangerous to do the autopsy. The body had to be carefully sealed in plastic and then sealed again in a metal container before it could be sent for study by scientists at University of California at Los Angeles.

Because of their potency and hardy life, anthrax spores have been used in biological warfare. The spores were expected to be used in biological warfare when Scandinavia supplied the Germans with anthrax in 1916, and the British experimented with it for use during World War II (1939–45). One plan involved creating "cattle cakes" injected with anthrax that would be dropped on Germany. (This never occurred.) Because of concern over anthrax being used in bioterrorism, American and British army personnel are routinely vaccinated against anthrax prior to serving in certain parts of the world. The vaccine that is used is 93 percent effective.

Shortly after the terror attacks of September 11, 2001, several letters containing a few grams of concentrated anthrax were mailed through the U.S. postal system, exposing people to anthrax. Mailed to several media offices and two Democratic U.S. senators, five people were killed and 17 others became ill from the exposure. In order to avoid further human contamination, the buildings where the letters were sent had to be thoroughly cleaned. Though better methods have since been devised, clearing the Senate Office Building of spores cost $27 million.

A firm identification of a suspect took a long time. After pursuing one particular scientist for a very long time, in 2008 the government's focus finally shifted. Strains of anthrax are unique, so the FBI examined the laboratories with the same strain of anthrax that was sent through the mail. Their continued inquiry led them to Dr. Bruce Edwards Ivins, a scientist who worked in a government biodefense lab at Fort Detrick, Maryland, where he had access to this particular type of anthrax. Shortly after being notified of the current line of investigation, he died of an overdose of Tylenol with codeine. There was no suicide note to verify suspicions, but law enforcement personnel feel that he took his own life

(continues)

(continued)

to avoid the consequences. In a January 3, 2009, article in the *New York Times,* Brad Garrett, a respected retired F.B.I. agent who had worked on the case, was quoted as saying that both "logic and evidence point to Dr. Ivins as the most likely perpetrator."

Today, scientists know that the best way to deal with anthrax used in bioterrorism is to come up with a way to detect it before people are exposed. In response to the October 2001 attacks, the United States Postal Service installed BioDetection Systems (BDS) in their largest mail cancellation facilities. In addition, community plans were drawn up for local responders to show them how to handle a situation where there was an indication that anthrax had been released.

Though experts still worry about anthrax, the reality is that it requires a relatively high level of expertise to make in the large quantities that would be suitable for warfare or any large-scale attack. A great deal of knowledge, training, and equipment are needed, and while it is certainly possible, it is not a first-choice option for most U.S. enemies.

(continued from page 9)

Using Pasteur's theory and Koch's postulates, scientists began to figure out cures for disease after disease. Pasteur's germ theory became the foundation of the science of microbiology and a cornerstone of modern medicine. Koch went on to discover the cholera bacillus (1892) and also the cause of tuberculosis (TB), though he was unable to determine a cure. In 1905, Koch received the Nobel Prize in medicine or physiology, primarily for his work on the causes of TB.

Koch also made another lasting contribution to scientific study, one that is still used today. He created pure methods for growing

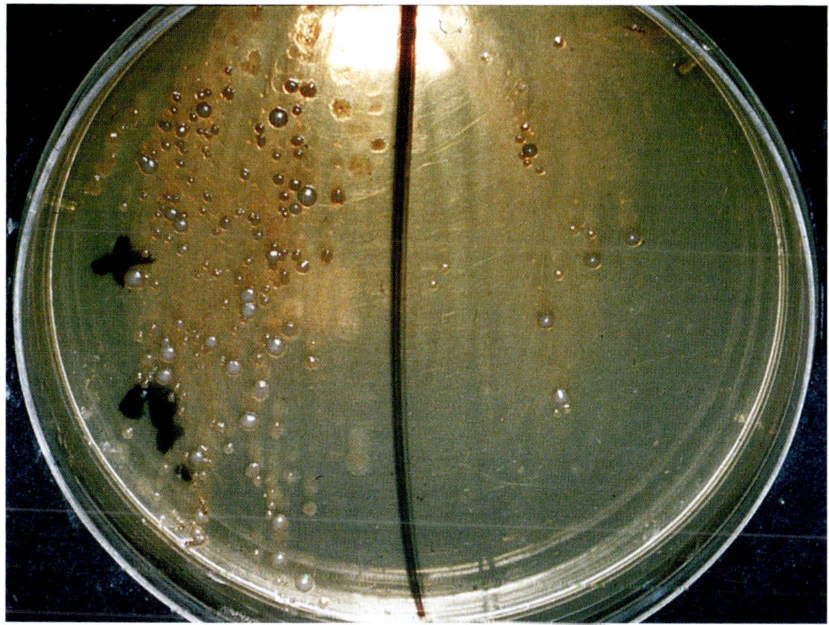

cultures of bacteria using a gelatinous medium called agar, which is composed from seaweed. The culture dish he used was invented by his assistant Julius Richard Petri.

IGNAZ SEMMELWEIS (1818–1865): IDENTIFIES THE CAUSE OF HOSPITAL INFECTIONS

Medical knowledge in the mid-19th century was desperately inadequate. In hospitals, surgery was performed without gloves and instruments were wiped clean on the physicians' aprons. In the 1840s, Ignaz Semmelweis was a successful obstetrician at the Allgemeine Krankenhaus (Vienna). At the hospital, it was common practice for doctors to do autopsies in the morning and perform pelvic examinations on expectant women or deliver babies in the afternoon. No one knew about sterilization of instruments or the importance of washing hands or wearing gloves, and puerperal fever (childbirth fever) was rampant.

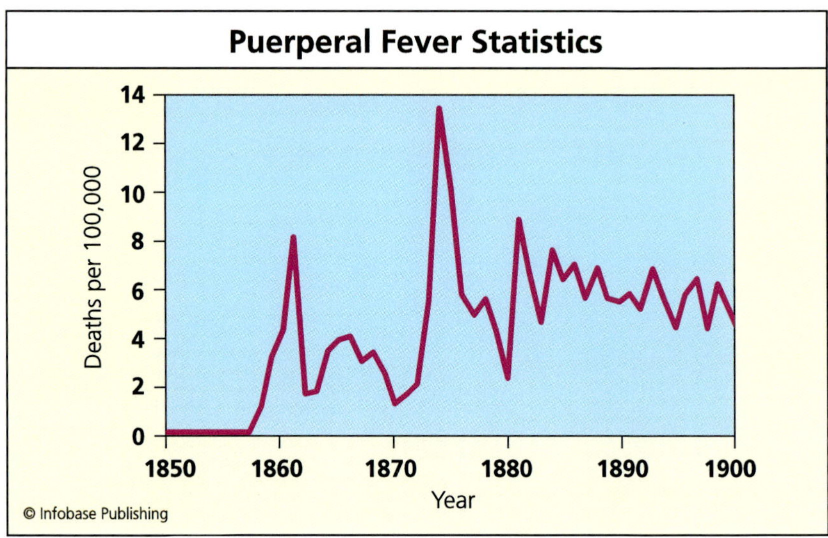

Ignaz Semmelweis was the first surgeon to make the connection between unclean surgical practices and the occurrence of puerperal fever. Semmelweis was unable to persuade many to change their ways. In the 1870s, Joseph Lister began to teach about cleanliness, and in 1878 Robert Koch demonstrated that surgical tools could be sterilized using steam. All of these elements helped reduce infection.

Dr. Semmelweis thought lack of cleanliness might be to blame for the high rate of the illness among new mothers, and he ordered the doctors to wash the pus, blood, and tissue from their hands after the autopsies and before seeing patients. Deaths from infections on Semmelweis's wards plunged (from 12 percent to 1 percent), but because Semmelweis had an an abrasive personality and could not communicate his ideas well, the hospital staff was resistant to his ideas for long-term change.

A few years later, Joseph Lister (1827–1912), a Scottish physician, rediscovered the importance of cleanliness in surgical proceedings and was able to bring about change.

Antiseptic Methods Adopted

By the middle of the 1800s, postoperative sepsis infections accounted for the death of almost half of the patients undergoing major surgery. A chemist by the name of Justin von Liebig determined that sepsis occurred when the injury was exposed to air, so starting in 1839 he advocated that wounds should be covered with plasters. Lister was skeptical of this explanation and von Liebig's recommendation. Lister had devoted a good number of years studying inflammation of wounds at the Glasgow Royal Infirmary and eventually was promoted to be the chief surgeon there, which provided an opportunity to take a look at the overall picture of processes at the infirmary. One of the facts Lister noted was that 45 to 50 percent of the amputation cases in the male accident ward were dying of sepsis (1861–65).

Lister suspected that a cleaner environment might be helpful. He began wearing clean clothes when he performed surgery. (This was not the norm for the day—surgeons frequently considered it a badge of honor to appear in blood-spattered aprons.) He also washed his hands before each procedure. At first Lister made no noticeable progress.

Then he became aware of the work being done by Louis Pasteur. Pasteur's work suggested that decay came from living organisms that affected human tissues, and Pasteur advocated the use of heat or chemicals to destroy the microorganisms. Lister determined that Pasteur's microorganisms might be causing the gangrene that so often plagued surgery patients and decided that chemicals would be the best way to stem the spread of microorganisms during and after surgery. He read that carbolic acid was being used to treat sewage in some places, so he created a solution of carbolic acid and began to spray surgical tools, surfaces, and even surgical incisions with his newly created mixture. For the next nine months, his patients at the Glasgow Royal Infirmary remained clear of sepsis.

At first, London and the United States resisted this theory; though they quibbled less about the theory of germs, they disagreed with the use of carbolic acid. To overcome this resistance, Lister arrived to become chair of clinical surgery at King's College where he began performing surgery under *antiseptic* conditions. Without much delay, his methods were accepted. Within just a few years, other surgeons began using Lister's antiseptic methods, and in 1878 Robert Koch demonstrated that steam could be used for sterilizing surgical tools and dressings. While the methods of sterilization have changed over the years, the concept of antiseptic surgery is still vital to success.

RUDOLF VIRCHOW'S CELLULAR DISCOVERIES

Rudolf Ludwig Karl Virchow (1821–1902) is known as the founder of cellular pathology because of his extensive research that stated that disease is created and reproduced at the cellular level of the body. While his discovery preceded the work of Louis Pasteur and Robert Koch and would have affected their thinking, Virchow's discovery took a long time to have any effect on patient care. However, his work created a foundation for a vital part of modern medical science.

From the early 17th century when scientists started peering through microscopes, they were fascinated by being able to view a world they could not see with normal vision. Many spent time investigating and theorizing about what they were seeing, and two scientists preceded Rudolf Virchow in noting the existence of cells in their different fields of study. The German botanist Matthias Jakob Schleiden (1804–81) was the first to recognize that all plants, and all the different parts of a plant, are composed of cells. Schleiden was friendly with the zoologist Theodor Schwann (1810–82) and mentioned to Schwann what he had observed in his plant studies. Schwann took a new look at the animal tissues he studied and realized that plants and animals seemed to share this commonality. This was quite a new thought in science. In 1839, Schwann was the first to write about cell theory when he published "Microscopic Investigations on the Accordance in the Structure and Growth of Plants and Animals."

Rudolf Virchow was familiar with the work of Matthias Schleiden and Theodor Schwann. Almost 20 years later (1858), Virchow defied many scientists of the time by teaching "Omnis cellula e cellula" or "Every cell originates from another cell." (Some scientists continued to believe that all matter was generated spontaneously.) Virchow published *Cellular Pathology* in 1858, where he addressed his reasoning that diseases also begin at a cellular level. This was a revolutionary thought for the time, and in his teachings he always encouraged students to "think microscopically."

Virchow's Earlier Life and Other Contributions

Virchow was born into a farming family and studied medicine on scholarship, gaining a medical degree in 1843 at the University of Berlin. In 1848, he was sent to investigate a typhus epidemic in Upper Silesia (part of what is now Poland), and in his report he stated that such outbreaks were caused not merely by poor hygiene but by conditions that a better government could help rectify—poverty, illiteracy, and political subjugation. (The Prussian government was busy dealing with a revolution in Berlin so they did not go after him, but his outspokenness did cost him an early professorship.)

From this time forward, Virchow became very active in campaigning for better standards for public health to help control the spread of illness. From 1859 to 1893, he served on the Berlin city council where he argued for inspection of meat and poultry, and he designed a plan for modern sewage disposal in the city. During the Franco-German War, Virchow helped train workers and provided medical care for soldiers.

Autopsy Findings

One of his greatest accomplishments happened much later in his career. During his lifetime, Virchow had spent a great deal of time in the laboratory, and much of what he had learned he taught himself by doing autopsies. By 1874, his organized and methodical system had become well known and other physicians came to learn his technique. Virchow's system is still one of the two methods used in autopsies today and involves removing each organ one by one. Others had advocated organ removal in units.

As a result of increasing autopsies, academic institutions began to create pathology departments to study the diseased tissues and body parts as they were removed. This created a new focus for science. Physicians began cataloging their findings, and while much

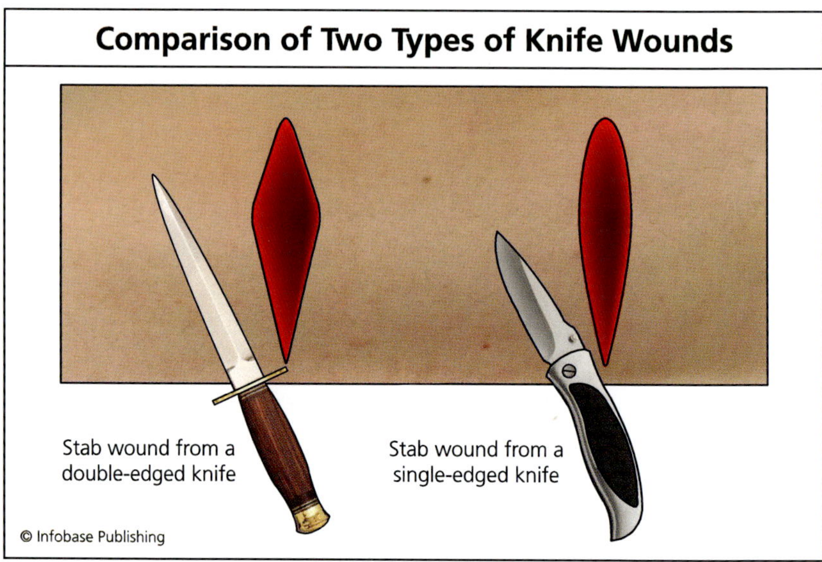

Physicians were beginning to accept autopsies as a tool in understanding death and disease. While today's medical examiners can use body temperature, blood analysis, and stomach contents and other tools to provide a detailed story of how and when someone died, when autopsies were first conducted scientists could really only observe the most superficial things, such as the type of instrument that might have left a particular wound.

of it was meaningless at the time, it provided vital information for scientists as study continued.

In addition to hospital autopsies to learn more about underlying diseases, the 19th century also saw an increase in the number of autopsies being conducted as part of criminal investigations. While their studies were extremely primitive compared to crime scene investigators' work today, scientists began to understand the differences in types of surface wounds and other causes of death.

CONCLUSION

The mid-19th century was a time of robust accomplishment. Virchow's identification of the importance of the cell, Pasteur and Koch's work on germ theory as well as on practical solutions to daily

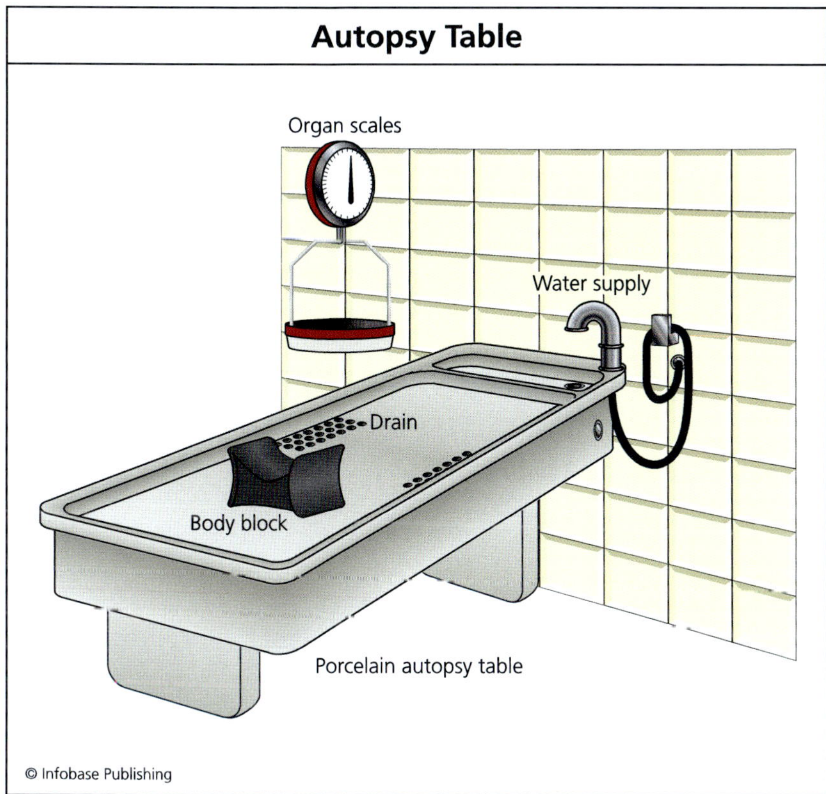

While several methods of conducting an autopsy are used today, one of the most commonly used is the one devised by Rudolf Virchow, who taught that organs should be removed one at a time and studied separately.

problems like the spoiling of wine and the spread of anthrax brought science into the mainstream. Businesses and government saw that there were very practical ways to benefit from the new theories.

Ignaz Semmelweis and Joseph Lister together ushered in a new era of hospital management that greatly reduced the infection rate and highlighted the need for sanitation as part of any medical treatment process.

2

Women and Modern Medicine

Women have always played an important role in health care, and, although there were female physicians in some cultures in ancient Egypt and in the early Middle Ages, women were relegated to serving as local healers, family caregivers, and midwives. With the exception of midwifery, their jobs were positions where they "learned by doing" and were often self-taught.

In the 19th century, these circumstances began to change. A very few women graduated from medical schools and became physicians. Women like Elizabeth Blackwell not only broke barriers for other women to become doctors, but many of them extended themselves by establishing organizations that made it easier for other women to follow.

Other women helped professionalize the field of nursing so that there was organization and a methodology to what was done. Florence Nightingale led the way for nurses to be an integral part of wartime medical care, and the methods she used when setting up her hospitals in the Crimean area served her well when she returned to England. The lesser-known Mary Seacole also made great contributions to nursing during the same period as Nightingale, but Seacole's contributions were not acknowledged for a long time because of racial prejudice.

Clara Barton not only worked in nursing, but she saw the need for the United States to join an international movement now known as the Red Cross. The ideals that guided this organization in the beginning are still in place today. Its original intention was to provide "disaster relief without prejudice." Today, most people would say they do even more.

Dorothea Dix observed the inhumane treatment of those with mental problems and dedicated her life to improving the care of those who could not speak for themselves. Alice Hamilton created a new field, industrial toxicology, when she realized the health problems that resulted from unclean and unsupervised working environments.

These are just a few of the women who have contributed greatly to medical progress, but their experiences are highlighted here because they demonstrate women who overcome the hurdles that were often placed in their paths. Each of these women contributed significantly to important steps forward in medical care.

ELIZABETH BLACKWELL (1821–1910): FIRST WOMAN DOCTOR IN MODERN TIMES

While Elizabeth Blackwell's name will always be connected with her status as the "first woman doctor," she also contributed significantly to changes in the medical profession. Blackwell established a hospital in New York City where poor women and children could come for care, and during the Civil War she and her sister founded the Woman's Central Association of Relief that was a vital part of providing better care for all soldiers, particularly the wounded, during the war. Later on, she opened a Woman's Medical College in New York to offer easier access to education for women.

Blackwell was born in England to a Quaker family. British custom dictated that upper-class women were only supposed to marry well, but the Quakers were more broad-minded. Blackwell's

Elizabeth Blackwell *(National Library of Medicine)*

parents felt strongly that their daughters should receive the same education as their sons. A spirit of social justice pervaded the family; they were also strongly antislavery and two of her brothers, Henry and Sam, married *suffragettes* (Lucy Stone and Antoinette Brown, respectively).

In her teens, Blackwell moved with her family to America, and they eventually settled in Cincinnati. Her father died soon after their arrival. To support the family, her mother established a school where she and her daughters taught. During this time, Blackwell nursed a good friend who was dying, and her interest in becoming a physician is thought to have been inspired by this experience. The woman complained of rough treatment by the male doctor, and she encouraged Blackwell to become a physician. Blackwell decided to do what she could to be a doctor, and she convinced two family friends who were physicians to let her read (study) under them while she continued to teach and save money.

Undaunted by the fact that no woman had ever been admitted to an American medical school, she applied to 30 schools and was rejected by 29 of them before she received an acceptance letter from Geneva College (now Hobart and William Smith Colleges in Geneva, New York). Her admission had been intended as a joke, but Blackwell had no way of knowing this. She arrived when classes started, took what she felt was her rightful place, and began her studies. The school administration decided to let her stay although they barred her from attending classes on topics that might have led to "embarrassing" discussions. In January

1849, she received her diploma, and many women in the community attended the graduation ceremony to signal their support of her.

Blackwell was now a naturalized U.S. citizen, but she felt her next step should be to return to Europe to learn more about medicine. The only hospital that granted her access was La Maternité de Paris, the lying-in hospital for poor women, which had a midwife-training program that required no prior education. Though she was treated in the same manner as the uneducated French girls, Blackwell felt she was learning, so she stayed. She eventually moved to London to study at a hospital there, but she contracted an infectious eye disease and lost one of her eyes as a result.

In 1851, she returned to the United States where she attempted to start a private practice, but paying patients were not interested in going to a woman doctor, and she had very little business. This experience was to be a pivotal one for Blackwell. Though it took time for her to gain adequate funding, she did so, establishing the New York Infirmary for Indigent Women and Children in 1857. (The hospital still exists; it is located in lower Manhattan and is now known as New York Downtown Hospital, owned by New York University.) She also hired the second woman to earn a medical degree in the United States, Marie Zakrzewska, a German-born physician of Polish descent. (Zakrzewska went on to found the New England Hospital for Women and Children, the first hospital in Boston.) Blackwell's younger sister Emily, who also became a physician, soon joined them. Blackwell focused on two particular missions with her hospital. By accepting women to study and work at her hospital as both nurses and physicians, Blackwell opened more educational opportunities for women. She also felt strongly about teaching both laypeople and professionals about the importance of hygiene and preventive medicine.

During the Civil War, Elizabeth and Emily Blackwell founded the Woman's Central Association of Relief. The original intent of the organization was to hire and train nurses for war service. Over time, it evolved into the United States Sanitary Commission, the federal agency responsible for training nurses and coordinating volunteer efforts. The Commission also provided battlefront hospital and kitchen services.

After the war, in 1868, Elizabeth and Emily opened a Woman's Medical College next to the hospital they had founded (with support from Florence Nightingale, discussed later in this chapter). Shortly after, Elizabeth was summoned back to London to "do for the British what she had done for American women." She spent the remainder of her life in London and cofounded the London School of Medicine.

Blackwell wrote several books that primarily had to do with women and medicine:

- *Lectures on the Laws of Life* (1852): a book that presents the case for physical education and exercise for children—a lifelong cause for Blackwell. It was written simply enough that it was accessible to all readers, not just those in the medical profession.
- *Medicine as a Profession for Women* (1860): this book was cowritten with her sister Emily and advocated for more women to enter the medical profession.
- *Pioneer Work in Opening the Medical Profession to Women* (1895): her autobiography.

Blackwell also wrote several other books, including one on parents' responsibility to provide strong moral values for their children, and another about sex, a topic not often addressed in the 19th century. Blackwell's contributions to medicine far exceeded the act of breaking barriers to become a doctor, because she made certain to turn and help those who might follow her. By the time she died in 1910, more than 7,000 women in America had become physicians.

THE PROFESSION OF NURSING

Though there have always been women tending to the ill, the profession of nursing actually has a relatively brief history. While wives and mothers throughout time have provided nursing care within a family and even a neighborhood, any organized care within a community was generally done by a religious order that dedicated itself to caring for the sick. If soldiers on battlefields were in need of medical or nursing care, that care was generally provided by a fellow soldier.

The primary role of nurses is sometimes assumed to be to carry out the instructions left them by physicians, but this is often not the case. When nursing first began during the mid-19th century, there were not enough physicians to handle the wartime patient load, so nurses who tended to the injured were often the best and only medical care a soldier would receive.

Treatment of the whole patient is the core belief of the nursing profession. While some medical specialties manage just one aspect of a patient's condition, the profession of nursing prides itself on a holistic approach. (Economic realities and too many patients sometimes prevent this from being a reality, but it is still the intent of those who go into the profession.)

Today, there has been an additional shift in the profession, and nurses care for all types of people suffering all types of illnesses—with nursing specialties ranging from mental and physical to neonatal and specialists in anesthesia. Degrees range from licensed practical nurse (LPN) to registered nurse (RN). The designation of nurse practitioner (NP) is for a registered nurse who has completed specific advanced nursing education (generally a master's degree) and trained in the diagnosis and management of common as well as complex medical conditions. In many states, nurse practitioners are considered qualified to provide basic medical care on their own, diagnosing, treating, and prescribing without having to work under a physician.

NIGHTINGALE AND SEACOLE: THE WOMEN BEHIND THE MOVEMENT

Florence Nightingale and Mary Seacole were two women at the forefront of professionalizing nursing care. Both got their start during the Crimean War. Nightingale has long been acknowledged for her contributions, but Mary Seacole, who was part Jamaican, faced racial discrimination, and this kept her work from being acknowledged for a very long time.

Florence Nightingale (1820–1910): Lady with the Lamp

Florence Nightingale laid the foundation for the nursing profession when she set out to improve the conditions for soldiers during the Crimean War in the 1850s. She worked tirelessly to implement hospital reform and bring compassion to patient care. Nightingale also developed a way to collect data and systematize recordkeeping of patient care, something that was not done regularly at the time. Her efforts proved the benefits of maintaining health statistics because she could graph a statistical report on disease trends or on how well a hospital was doing. This work was acknowledged when she became the first woman elected to the Royal Statistical Society (1859).

Florence Nightingale was born in 1820 to a wealthy British family, and her childhood was spent at an estate in Derbyshire, England. Well-to-do young women of the time were trained to be refined ladies so they could marry well, but William Edward Nightingale believed his daughters should receive the same type of education as boys. Florence and her sister were taught Italian, Latin, Greek, history, and mathematics. Florence particularly excelled at mathematics, and her father was happy to teach her all he could.

Though there was a sentiment at this time that the sight of naked flesh would corrupt young women, Nightingale volunteered at area hospitals and felt she was answering a divine calling, so she worked out a way to gain the necessary knowledge. In 1846, she visited Kaiserwerth, a pioneering hospital in Germany established and managed by an order of Catholic sisters, and she was greatly impressed by the quality of medical care and by the commitment and practices of the sisters. Later, Nightingale returned to Germany and spent four months studying to be a nurse at Kaiserwerth.

Her next steps would not have been possible had Nightingale not had the necessary political connections through her family.

The Lady with the Lamp from a painting by Honrietta Rae (Library of Congress Prints and Photographs Division)

When the Crimean War broke out in 1853, France's soldiers were aided by women from several religious orders, but British medical care was seriously lacking. With the approval of her friend Sidney Herbert, who held a governmental office, Nightingale selected and trained 38 volunteer nurses whom she took with her to the

Crimean area. The group arrived in Scutari (now part of Istanbul) and found desperate circumstances. The temporary hospitals were unclean, and there was little equipment to use to help care for the patients. The injured were left lying in their filthy, blood-stained uniforms, and there was no soap or towels or clean clothing. There was very little to eat and no containers to take water to the men. Shortly after her arrival, the death rate actually rose and was the highest of any hospital in the area. Nightingale saw that 10 times more soldiers were dying from illnesses than from their wounds. She contacted the British government and implored them to send help. A sanitary commission was deployed; they flushed the sewers, helped obtain freshwater for the hospital, and improved the ventilation. Nightingale kept careful statistical records of how patients at the hospital were faring, and after the sanitary commission brought about the needed changes she was able to prove that the death rate dropped from 42.7 to 2.2 percent. Her point was made.

Nightingale came to be known as a sign of hope, the lady with the lamp, partly because of the good work she did for the soldiers, but also because she was the last person to go through the hospital each night to check on all of the patients—she really was the lady with the lamp.

Her good deeds quickly became known in Britain, and when she returned home a fund was started to recognize her for her work. Nightingale directed that the money be used for setting up the Nightingale Training School in 1860 (now known as the Florence Nightingale School of Nursing and Midwifery, a part of King's College, London). In 1860, she also wrote and published *Notes on Nursing,* which is considered a classic introduction to nursing. (It was also simply written and the home nursing advice was taken up by many laywomen who read it to learn more about providing care at home.) Later she wrote *Notes on Hospitals* (1863).

Mary Seacole (1805–1881): Fought Barriers to Help

Mary Seacole was a multiracial woman born to a Scottish father and a free black mother. Her mother was a "doctress" [sic], and the family ran a boardinghouse in Jamaica where those suffering from tropical diseases often stayed until they got better. Seacole learned about patient care from her mother and spent several years in Central America and the Caribbean where she became familiar with cholera and other illnesses that were common in the Tropics.

When the Crimean War broke out, Seacole heard of the intended use of nurses near the battlefield, and she wanted to help. Armed with letters of recommendations from area doctors, she traveled to London where she attempted to meet with someone so that she could volunteer. However, no one took her up on her offer nor was she selected to be among Nightingale's volunteers. Seacole assumed personal responsibility for getting to the Crimea and set about raising money for her travel expenses. Once in Crimea, she was again turned away by Florence Nightingale, so she established a hotel in the area—probably not unlike the boardinghouse her family had run in Kingston, Jamaica where she took care of the sick and wounded. Nightingale was dismissive of Seacole's efforts. Because Seacole's retreat was also a hotel, alcohol was served, and this led Nightingale to refer to it as little more than a brothel.

Though Seacole was long overshadowed by Nightingale's contributions to nursing, the 21st century has seen a new understanding of the racial obstacles that stood in Seacole's way. Several nursing prizes have been established in her honor, and a long-running exhibit of her contributions was held at the Florence Nightingale Museum in London to celebrate her very real contributions to nursing.

CLARA BARTON (1821–1912): FOUNDER OF THE AMERICAN RED CROSS

Clara (Clarissa Harlowe) Barton was a dedicated humanitarian who recognized a need—that of taking supplies to the soldiers in the field during the Civil War—and stepped in to help out. She was nearly 40 years old when she started traveling with wagons to the battlefront to provide medical supplies and food

LINDA RICHARDS (1841–1930):
First Professional Nurse in the United States

Linda Richards was the first professionally trained American nurse. She is credited with establishing nurse-training programs in various parts of the United States and in Japan. She also is recognized for creating the first system for keeping individual medical records for hospitalized patients.

Richards's early life experiences directed her toward an interest in nursing. While she was still a child, both of her parents died of tuberculosis. She was four when her father died but a young teen when her mother became ill, and she nursed her mother until the end of the illness. Living in Newburyport, Vermont, there was no way to prepare for becoming a nurse, so Richards trained to be a teacher and soon married a local farmer who then went off to serve as part of the Green Mountain Boys (a unit of Vermonters) in the Civil War. In 1865, her husband returned, wounded, and Richards nursed him until his death in 1869.

where they were most needed. Later, at the age of 60 (1881), she founded the American Red Cross and led it for the next 23 years.

Clara Barton was working as a clerk in the U.S. Patent Office in Washington, D.C., when the Civil War started. The Sixth Massachusetts Infantry had been attacked in Baltimore, Maryland, by southern sympathizers. The men were temporarily housed in Washington in the unfinished Capitol building. Barton appealed to the public for donations, gathered items herself, and also collected relief supplies from the U.S. Sanitary Commission. Barton also offered personal support to the men in hopes of keeping their

Richards moved to Boston and took a job at Boston City Hospital, but she was relegated to cleaning chores. She soon heard of an educational program started by Marie Zakrzewska and was one of five women to sign up for the nurse-training course at the New England Hospital for Women and Children. In 1873, she was the program's first graduate.

She eventually moved to New York City where she became the night supervisor at Bellevue Hospital and created a system for keeping individual records for each patient. Her system became widely used in this country and in England. As her career progressed, Richards became intent on establishing more nurse-training programs. Her program in Boston became known as one of the best in the country. She went on to establish and direct nurse-training programs in Pennsylvania, Massachusetts, and Michigan. She also traveled to Japan and established the first training program for nurses in that country (1885–86).

spirits up: She read to them, wrote letters for them, listened to their personal problems, and prayed with them.

Though it was highly unusual, Barton wanted to follow the men to the front lines and, after much effort, Barton was eventually given passes to bring her voluntary services and medical supplies to the battlefront and to field hospitals. Her first trip was to Virginia in August 1862, and, when she arrived with her supplies, the overwhelmed surgeon on duty wrote later, "I thought that night if heaven ever sent out a[n] . . . angel, she must be one—her assistance was so timely." She became known as the Angel of the Battlefield.

Clara Barton *(Library of Congress Prints and Photographs Division)*

After the war, Clara Barton visited Europe in 1869 and was introduced to a book by Henry Dunant, who had founded the Red Cross movement. Henry Dunant (1828–1910) was the son of a Swiss businessman who witnessed horrific fighting in 1859 at the Battle of Solferino in Italy. This experience led him to social activism. He devised the idea for an International Red Cross that he intended as a multicountry movement to protect the sick and wounded during wartime without respect to nationality. As part of his dedication to getting this idea accepted, Dunant pushed hard for what became known as the Geneva Conventions at which a treaty embodying Dunant's idea was negotiated. This treaty (also referred to as the Geneva Treaty or the Red Cross Treaty) was ratified by 12 European nations in 1864, and Clara Barton campaigned tirelessly, and ultimately successfully, for the United States to ratify it as well, which the country ultimately did in 1882.

In 1881, Barton formed the American Association of the Red Cross. In 1893, it was reincorporated as the American National Red Cross, and it received charters by Congress in 1900 and 1905 that provide for a close working relationship with the government. The first time Barton's organization provided aid was in 1881 to victims of a devastating forest fire in Michigan; in 1884, she chartered steamers to carry needed supplies up and down the Ohio and Mississippi Rivers to assist flood victims. In 1889, she and 50 volunteers rode the first train into Johnstown, Pennsylvania, to help the survivors of a dam break that caused 2,200 deaths. While disaster relief remains a primary mission of the organization, the Red Cross also sends help to war zones; 1892 was the first time that the American Red Cross provided assistance to American armed forces and civilians during wartime.

DOROTHEA DIX (1802–1887): SOCIAL REFORMER AND ADVOCATE FOR THE INSANE

At a time when no one understood the causes or issues surrounding mental illness, Dorothea Dix gave voice to those who could not advocate for themselves, the mentally ill. Later in life she served as superintendent of female nurses for the Union during the Civil War, an unpaid position for which she volunteered.

Dorothea Dix was born in Maine to an itinerant preacher, and her early intention was to become a teacher. After receiving her education, she approached her well-to-do grandmother about funding a school to offer opportunities to more girls to get an education, and her grandmother provided Dix with space for the school in her home in Boston.

In her mid-30s, Dix suffered an emotional breakdown. She went to England to recover and was befriended by the Rathbones, a Quaker family dedicated to social reform. The Rathbones were very involved in a lunacy reform movement in Britain that was dedicated to investigating madhouses and asylums, and this was to have a major effect on Dix's life.

When Dix returned to the United States in 1840–41, she saw firsthand what

Dorothea Dix *(Library of Congress Prints and Photographs Division)*

was happening to the mentally disadvantaged in her own country. She volunteered to teach a Sunday school class for women inmates at the East Cambridge jail, and this gave her the opportunity to see the terrible living conditions of the prisoners. She also noted that prostitutes, drunks, criminals, retarded individuals and the mentally ill were all housed together. When she asked why the mentally ill were being kept in jail, she was told "the insane do not feel heat or cold."

This experience inspired her to begin an investigation of how the state of Massachusetts cared for the insane poor. She found that, typically, towns paid local individuals to care for people with mental disorders. However, the system was unregulated and under-funded, and it produced widespread abuse. According to a report presented by Dix to the state legislature, people were kept in cages, in stalls, and in pens, while others went naked, and chained in place. Many were beaten to make them obedient. As a result of Dix's efforts, Massachusetts finally set aside money to expand the state's mental hospital in Worcester to accommodate more patients. Dix then focused on other states, traveling from New Hampshire to Louisiana to document the conditions of the insane.

Dix's views were radical for the time. People believed that the insane would never be cured, so it did not matter in what way they were housed. Dix's actions led to better living conditions, and, in some cases, this led to improvements in patients' mental health as well. She played a major role in founding 32 mental hospitals, 15 schools for the "feebleminded," a school for the blind, and numerous training facilities for nurses.

When the Civil War began, Dix suggested that women could be recruited and trained to help in military hospitals, and she volunteered to help out. In June 1861, Dix was appointed to recruit and supervise these women. Dix's advocacy for women in these positions broke certain gender barriers, but she maintained certain prejudices. She didn't want marriage-minded young women working near the soldiers, so she insisted applicants be over 30 and "plain looking." She also implemented a dress code of black or brown skirts, no hoopskirts, and no jewelry.

Dix was better suited to be a social reformer than a member of a bureaucracy. She clashed frequently with army officers. Nonetheless, nursing care was much better under her leadership. She was capable of procuring badly needed medical supplies from private sources, and she enforced the need to care for southern soldiers as well as northern ones.

ALICE HAMILTON (1869–1970): STUDIED FIELD OF INDUSTRIAL TOXICOLOGY

Alice Hamilton became a chemist at a time when it was not easy for women to work in science, and she devoted her career to investigating workplace health hazards to which employees are exposed.

Hamilton was one of five children born to prominent parents in Fort Wayne, Indiana. She attended Miss Porter's School, a girls' boarding school in Farmington, Connecticut, intending to go on to medical school. (At that time, students could go from high school directly to medical school.) She began at the Fort Wayne College of Medicine in Fort Wayne, Indiana, and she continued her education at the University of Michigan Medical School, graduating in 1893. While in medical school she became fascinated by pathology so she decided to become a research scientist. Hamilton's sister was going to Europe to study the classics, and Hamilton decided to accompany her to continue her studies in bacteriology. (Her sister was Edith Hamilton, the world-renowned classicist whose *Mythology* remains most American children's introduction to mythology.) Universities in Munich and Leipzig had never before admitted female students, so she was permitted to attend lectures in bacteriology and pathology "if she made herself inconspicuous." She then returned to the United States where she became a researcher at Johns Hopkins Medical School.

In 1897, she became professor of pathology at the short-lived Women's Medical School at Northwestern University near Chicago. Hamilton attended a public lecture given by Jane Addams, the founder of the settlement house known as Hull House. Settlement houses were run by idealistic people who intended to provide badly needed housing for newly arrived urban immigrants. Hamilton decided that living at Hull House would provide insight into the health problems of the urban poor who seemed to have particularly high rates of typhoid fever and tuberculosis.

In 1902, Chicago had a typhoid epidemic, and Hamilton made a connection between poor sewage disposal and the role of flies in transmitting disease. Her information led to reorganization of the Chicago health department, and soon the governor of Illinois appointed her to the Illinois Commission on Occupational Diseases, which ran a groundbreaking study surveying industrial diseases in Illinois. At the time, there were no laws regulating safety in the workplace. Her efforts became well known, and she was soon called upon to address issues for the national government. From 1911 to 1920, she served as a special investigator for the federal Bureau of Labor (later called the Department of Labor) where she undertook a study of the use of lead and lead oxide in manufacturing paint. (No one understood the dangers of lead poisoning at this time.) She also noted the health problems of those who were exposed to noxious chemicals at work.

By 1919, Alice Hamilton was the acknowledged expert in the field of industrial medicine, and, although Harvard University's entire faculty was male, the medical school determined that they would create a department of industrial medicine with Hamilton as its leader. She was given the title of assistant professor. However, Harvard placed three qualifications on her appointment: Dr. Hamilton was not to use the Faculty Club; she would not be permitted to march in commencement processions with the rest of the faculty; and she would never be given football tickets, a perk available to the rest of the faculty. She took the position and was never given a higher standing than assistant professor.

Hamilton continued to devote six months a years to conducting her surveys of employment conditions within various industries, and over time she revealed to government and business the dangers of certain dyes, carbon monoxide, mercury, lead, radium (commonly used in wristwatch dials), benzene, the chemicals in storage batteries, and carbon disulfide and hydrogen sulfide bases which were created during the process of making rayon.

Hamilton dedicated her life to making life better for other people, and she was active with both health-related and political causes. She died in 1970 at the age of 101.

CONCLUSION

Today, women physicians are so common that it is shocking to think that only 160 years ago the women who entered the profession had a very difficult time. Elizabeth Blackwell was instrumental in helping other women enter the medical profession, and her own contributions to the field were enormous. The field of nursing was professionalized through the efforts of women like Florence Nightingale, Mary Seacole, and Linda Richards. Today, nurses can pursue a myriad of medical specialties, and there is a growing level of respect for the nursing profession. Dorothea Dix and Alice Hamilton both made vital contributions in fields that were getting little attention at the time. Dix addressed the needs of the mentally ill, and Alice Hamilton brought attention to the hazards of the workplace. Both these fields were to become increasingly important in the 20th century.

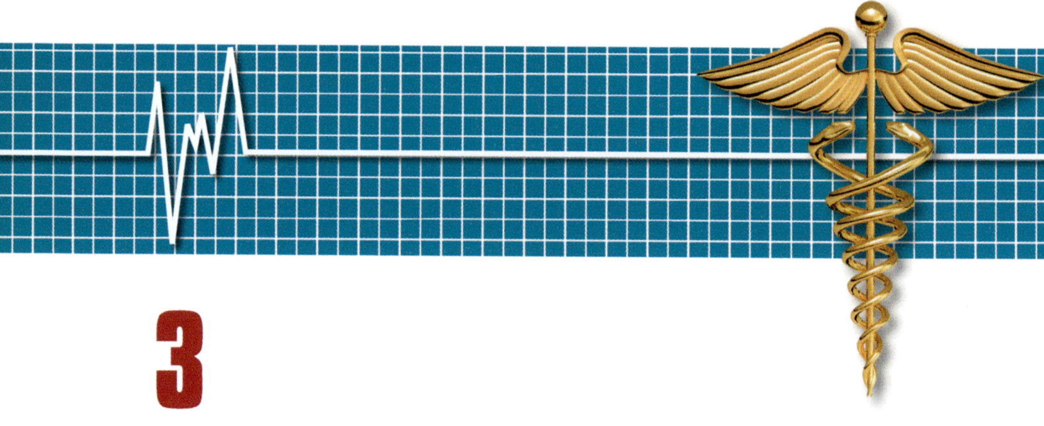

3

Science Moves Forward in Diagnosis and Treatment

By the mid-19th century, science was moving forward in many fields, and the information gleaned in various areas was proving useful to those looking for better methods of medical diagnosis and treatment. Several of the discoveries—anesthesia and *X-rays*—were accidental findings that proved highly beneficial. Others like the discovery of *viruses* came after a very prolonged period of study, and even then it was several decades before scientists had access to powerful enough microscopes to reveal the presence of the tiny infectious agents scientists had come to suspect were there.

Surgery was one treatment that was often necessary but extremely problematic. When other options were exhausted and surgery seemed the best recourse, doctors had no reliable way to dull the pain, and, after the surgery, there was no way to prevent infection. Most surgical procedures were limited to the extremities and superficial parts of the body, mainly because surgeons had limited understanding of how the internal organs worked. Surgeons were beginning to understand *anatomy,* but much of *physiology*—how the body worked—was still a mystery. The field of pathology was very young.

In the 1840s, two forms of anesthesia began to be used: one was *nitrous oxide* (often called laughing gas) and the other was ether (which became more popular for surgery). Nitrous oxide was first used with dental patients, and over time it became more commonly used by dentists than physicians. (Some dentists today still use it.) For surgical patients, ether was often administered by having a patient breathe into a cloth saturated with it, and this method of drug delivery was used into the 20th century. Because physicians were in disbelief that surgery could be accomplished without causing pain to the patient, these early operations with the patient under ether were frequently performed in front of audiences.

Diagnostics took a big jump forward with the introduction of the X-ray process. At the time X-rays were discovered, no scientist was searching for a method for "looking through" the human body, but once Wilhelm Röntgen made such a discovery, he and others quickly found ways to apply it to medicine. Marie Curie's work was helpful in many fields of science, but she took particular interest in helping people understand how radiation could be used diagnostically.

In other areas of diagnostics, scientists were quickly ticking off the identification of more and more types of bacteria that caused illnesses, but they were still baffled by the fact that they could not identify the causes of certain diseases. Ironically, physicians had created vaccines that could protect against diseases like smallpox, but they still could not identify the causative agent. Viruses were finally discovered after two scientists diligently studied diseased tobacco plants. With this new information, scientists were able to look for new types of cures. *Serum therapies* were tried based on what scientists were learning about creating vaccines, and this new method of treatment was introduced at this time.

This chapter reveals the surprising way that anesthesia was introduced for use on surgical patients, and it relates how the X-ray process developed. The chapter concludes with an explanation of why viruses were so difficult to identify and who it was who finally made the discovery of these pathogens.

THE EARLY USE OF ANESTHESIA

Through the early 19th century, patients had to endure great pain if they had surgery. Many years after the fact, one elderly Boston physician wrote of the horror of patients' yells and screams while he operated on them. The pain during surgery was so accepted—and so unavoidable—that some surgeons believed that keeping the patient conscious during surgery was actually beneficial and promoted healing. Surgeons used able-bodied assistants to help tie patients to the operating table for the procedure. Some surgeons looked for ways to dull a patient's sensations before surgery. A few physicians literally knocked their patients out with a blow to the jaw; others believed the solution lay in counterirritants, such as having an assistant rub another part of the patient's body with stinging nettles to distract the person from the part of the body undergoing surgery. A few doctors began experimenting with herbs and alcohol. Plants such as marijuana, belladonna, and jimsonweed were among those that provided some relief to patients prior to a procedure. Alcohol and opium also were used, but alcohol in large enough doses to produce a stupor tended to cause nausea, vomiting, and sometimes death instead of sleep. Opium had significant side effects as well, and typically it was not powerful enough to blunt the feeling of surgery.

The road to the development of anesthesia was built on the findings of a succession of people, and the discoveries started as early as the 17th century. The chemist and clergyman Joseph Priestley (1733–1804) noted and described the properties of nitrous oxide (laughing gas). Many years later, Humphry Davy (1778–1829) discovered the exhilarating effects of the gas and noted the possibility of using it for surgery. In 1818, the English chemist and physicist Michael Faraday (1791–1867) found that vapors from sulfuric ether also created mind-altering effects that could numb pain.

At this point, no one had employed ether or nitrous oxide for any practical use. However, the gases were popular among well-to-do partygoers and students who had access to laboratories. Those people who could obtain and bring along a bag of nitrous oxide or a flask of ether to share at a party were very popular, as everyone saw that parties were a lot more fun "under the influence." Even in a small town like Jefferson, Georgia, the fun of a party that included ether was something to relish. Dr. Crawford Long (1815–78), a physician, was accommodating fun-loving friends by providing ether for their parties when Long observed that those who had breathed the ether never seemed to notice if they tripped and fell or banged into things. This gave Long an idea. One of his friends, James Venable, had a tumor on his neck that needed to be removed. Long approached Venable and suggested that if Venable took a whiff of ether, perhaps Long could remove the tumor without causing too much pain. Venable's tumor was successfully excised, and Venable was not bothered by the process. Encouraged by this success, Long tested the ether when he performed minor operations on other people. However, it was a full seven years before he documented what he had learned, finally submitting an article on the process to *Southern Medical and Surgical Journal*. There was no adequate explanation as to why he waited so long to report his discovery, but some experts feel that since Long was a small-town doctor, he may have encountered a disapproving citizenry, and perhaps he worried that additional publicity for his use of ether would have negative effects on his practice.

In the meantime, ether frolics and nitrous oxide parties continued, and soon showmen realized they could benefit from the craze. Just as the men who demonstrated patent medicines created traveling road shows, other fellows developed shows involving nitrous oxide. Audiences paid to attend the events, and then lucky volunteers were invited to the stage to sniff the nitrous oxide; friends and neighbors were mightily entertained by watching what the volunteers did after breathing in the laughing gas. In 1844, a dentist by the name of Horace Wells (1815–48) attended a show in his hometown of Hartford, Connecticut. He and a companion willingly volunteered to participate, and afterward Wells noted that his friend had bumped into something and cut his leg while under the nitrous oxide and yet seemed to feel no pain. Wells thought this over and wondered about whether this gas could be helpful in dentistry, and he was in a perfect circumstance for finding out.

He himself had a wisdom tooth that was bothering him but he had resisted letting anyone pull it because of the pain. Wells solicited help from another dentist, and together they approached the fellow who ran the road show for help. Wells wanted to be given the nitrous oxide and then have his friend pull the tooth. The road show fellow agreed. Wells "went under," his friend pulled the tooth, and Wells did not feel anything.

Wells realized that to introduce the use of this gas to a wider population for general surgery was going to take some thought and effort. He got in touch with a friend and business partner, William Morton (1819–68), who had contacts in the Boston medical community. As Morton quickly saw this as an opportunity to profit and bring esteem to himself, he took Wells to meet the highly respected head of surgery at Massachusetts General Hospital, Dr. John Collins Warren. Surgeons were well aware that progress needed to be made in creating a way for patients to withstand surgery, so Warren thought this proposal was interesting. In January 1845, he arranged to pull a patient's tooth after the patient had breathed in nitrous oxide. Because this offered a groundbreaking opportunity, Wells arranged for the procedure to be conducted in front of an audience of interested medical professionals. Wells was asked to administer the appropriate dose of nitrous oxide, but when Warren pulled the fellow's tooth the patient cried out. Later the patient said he felt nothing and must have simply groaned, but the damage was done. Wells was humiliated, and the medical community continued its brutal but familiar methods for conducting surgery and dental work.

William Morton was not tainted by the experiment since he had merely provided an introduction to Dr. Warren, and he sensed there was still great opportunity. He began experimenting with ether, the other substance that was popular on the party circuit. In 1846, he got in touch with Charles Jackson (1805–80) a brilliant but reputedly cantankerous scientist. Morton consulted Jackson on how to use ether to numb a patient for dental work. Jackson provided Morton with some additional information, and Morton immediately set to work experimenting with it. Morton's work became known to another young surgeon who practiced at Massachusetts General, and, since the substance to be used was different, the surgeon was successful in encouraging Dr. Warren to give this new procedure a chance. Warren agreed and set an appointment for another public demonstration the next day.

Traditionally, doctors had always shared medical advances so that all could benefit, but Morton wanted to make the ether a *proprietary* product on which he could make money. He had to figure out a way to disguise what he was doing so that no one else could do it. With only 24 hours' notice of the demonstration, Morton created a slightly different delivery method and added orange oil to mask the basic ingredient he was using. He arrived for the surgery 25 minutes late, to the great annoyance of Dr. Warren who was about to go ahead and make the first incision. Warren reluctantly paused and permitted Morton to proceed with putting the patient to sleep—a feat that was accomplished quickly. The surgery was a success, and the patient felt no pain.

The use of ether for surgery spread quickly, and scientists began to experiment with other similar substances such as chloroform, which soon replaced ether in many operating rooms. Chloroform was easier to handle because it was less likely to combust, and patients tolerated it better. Ether often caused vomiting, and people who had chloroform were less likely to have an upset stomach afterward.

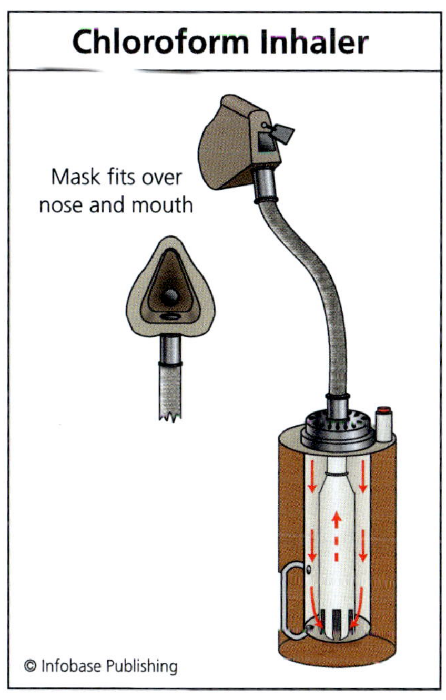

As the use of anesthesia grew, chloroform became more popular than ether for surgical procedures. Chloroform was easier to administer and caused fewer upset stomachs.

The use of chloroform before surgery grew quickly; it was a major advance for both patient and surgeon.

Today, anesthesia is administered by a specialist who has studied anesthesiology and not only understands the different types of anesthesia but how to carefully monitor the person during surgery. The anesthesiologist also helps determine whether general anesthesia—loss of consciousness—or local anesthesia, where the nerves in a particular part of the body are "frozen," is appropriate for each situation. Today, less than one patient out of 250,000 people die from anesthesia used for surgery. Interestingly, nitrous oxide has regained popularity in dentistry. It seems to be unique in having no toxic effects on the body.

While the gains made in anesthesia have been remarkable for patients, the end did not work out so well for those who introduced the process to the world. Crawford Long, who remained largely unrecognized for his work, is the only one who emerged unscathed from the experience. He continued to practice medicine until he died of a stroke in 1878. Wells switched from nitrous oxide to chloroform for his dental practice, and he soon became addicted to chloroform. He was arrested in 1848 for pouring sulfuric acid on a prostitute and committed suicide while in jail. Morton and Jackson ended their lives feuding over credit for creating a way to anesthetize surgical patients. Morton had gotten Jackson to cosign on a patent application for the use of ether. Morton called it Letheon, and he tried to sell licensing rights but the patent was worthless; others were readily able to figure out how to use ether for surgery so they saw no need to pay Morton for the rights to anything. Morton died at a young age, greatly frustrated. Dr. Jackson spent the rest of his life trying to be recognized for "inventing" anesthesia.

THE DEVELOPMENT OF X-RAYS

Wilhelm Conrad Röntgen was a dedicated physicist who made a major contribution to medical science by his accidental discovery of rays that were capable of passing through the human body. This discovery was to have a major effect on 20th-century medicine, and in 1901 Röntgen was awarded the Nobel Prize in physics for his 1895 discovery of X-rays.

In 1895, Röntgen was in his laboratory experimenting with electron beams emitted from a gas discharge tube (a Crookes tube). Röntgen noticed that a fluorescent screen in his lab started to emit a faint green glow when the electron beam was turned on. Fluorescent material was known to glow in reaction to electromagnetic radiation, so this would not have surprised him, but Röntgen's tube was wrapped with heavy black cardboard that should have blocked the rays. In order for the glow to be created, the rays had to have been passing through the cardboard. Röntgen tried placing books and papers between the tube and the screen, but the glow still occurred. For the next two months, he undertook systematic

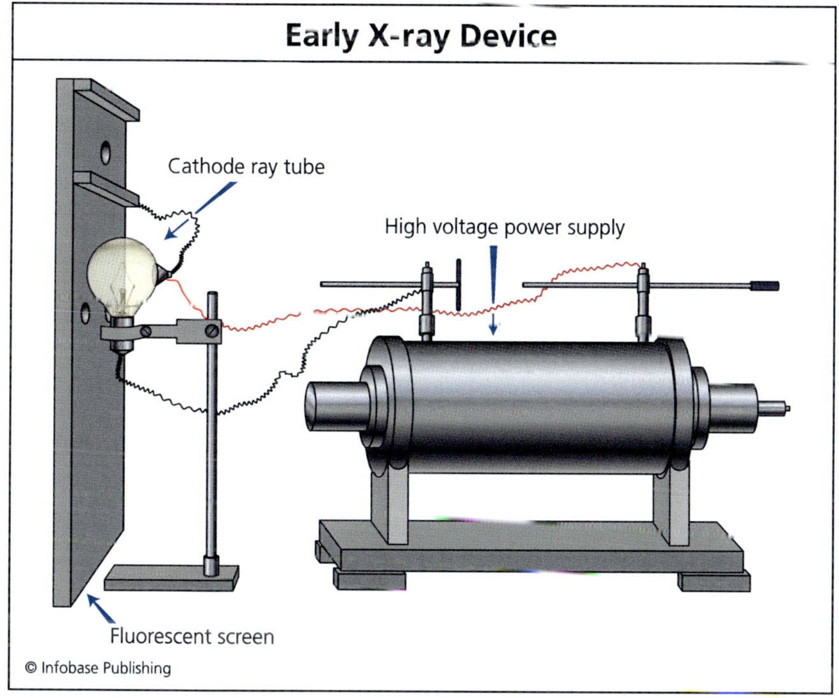

After Röntgen identified the meaning of his discovery of what he called X-rays, he created a device similar to this one so that he could further experiment with the process.

experimentation of what was happening. Because of the unknown nature of the rays, he called them X-rays.

As he experimented, he saw that the rays also passed through the human body, providing a silhouette of the bones within. To prove what he witnessed, Röntgen needed to capture it photographically. He asked his wife to let him photograph her hand as it looked when penetrated by radiation. This X-ray photograph was the first of a human body part. Upon seeing it, physicians began to realize how X-rays could be helpful in medicine. On December 28, 1895, he submitted a report "On a New Kind of Ray: A Preliminary Communication" to the Würzburg Physico-Medical Society. On January 5, 1896, a little more than a week later, a reporter for the Vienna press wrote a story, and excitement for the process began to spread around the world. Röntgen's remarkable discovery created a significant advance in diagnostic medicine, permitting doctors to see straight through human tissue. See the sidebar "The Application of Radiology" on page 50 for more information about the medical uses of radiation.

Unlike many medical discoveries, X-rays were embraced quickly. To people in the early 20th century, it may have seemed like what would be considered a science fiction advance today. To be able to view the skeleton through clothes—and flesh—seemed amazing. Within weeks, doctors were using X-rays to locate bullets and detect broken bones, and they soon found that if they had their patients drink barium salts dissolved in water, X-rays could reveal the esophagus, stomach, and small intestine. A solution of iodine could be used to diagnose problems in the bladder or kidneys, and other chemicals permitted doctors to see the veins.

Röntgen never tried to make money from his discovery, though others certainly capitalized on the public excitement over this new discovery. Thomas Edison built an X-ray device with a screen large enough to permit people to see their entire bodies via X-ray, and people lined up for turns seeing themselves. X-ray machines also began appearing at game arcades and in department stores. Shoe stores decided there was a very practical purpose for an X-ray machine for the foot that permitted shoe salesmen to check the fit

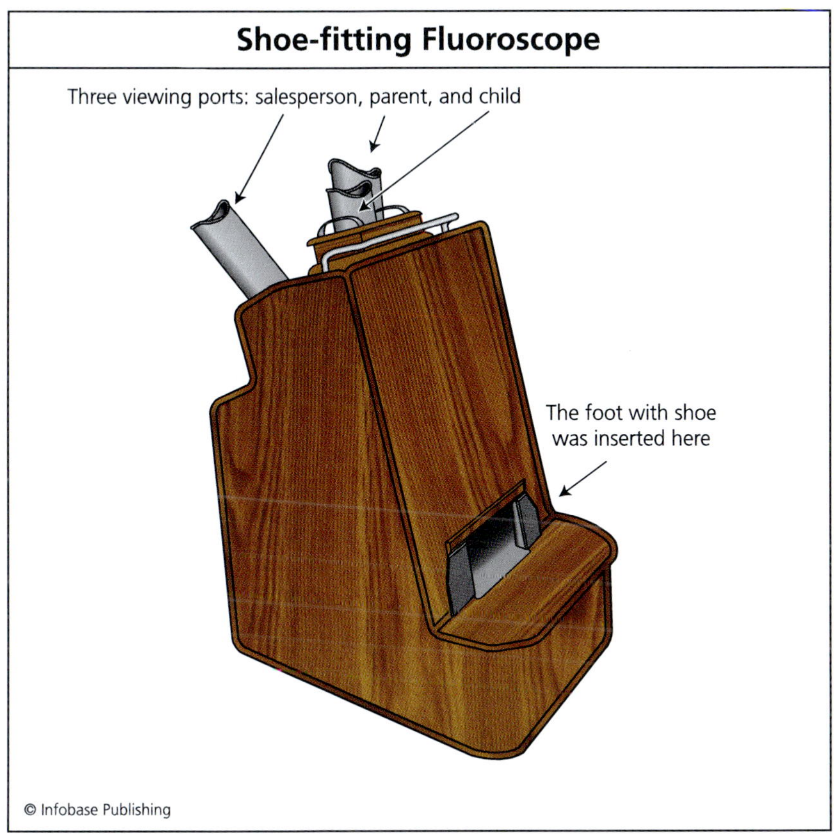

The ability to use radiation in many different ways became popular with the public. This is an example of a device that permitted a shoe salesman to see an X-ray of a child's foot in a new pair of shoes through the fluoroscope and then pronounce whether or not the shoe was the right fit.

of a shoe based on an X-ray. Unfortunately, the negative effects of X-rays soon became apparent, as those who worked near these new devices began to suffer from sickness caused by radiation exposure. Edison's assistant, Clarence Dally, was one of the first to become ill. His hands turned red, his skin peeled away, and his hair fell out, and he died within a few years. However, it was a long time before scientists started taking adequate measures to protect themselves when working with radiation.

MARIE CURIE (1867–1934): PIONEER IN THE FIELD OF RADIOACTIVITY

Marie Curie made major contributions to physics and chemistry and was the first person to win two Nobel prizes. She is remembered as the discoverer of the radioactive elements polonium and radium, and, while this discovery changed scientists understanding of matter and energy, the application of this knowledge also ushered in a new era in medical research and treatment.

Building on the work of Wilhelm Röntgen and his discovery of X-rays, Marie and Pierre Curie (Pierre was a physicist married to Marie) along with French physicist Antoine-Henri Becquerel began investigating what other elements might emit similar rays. Becquerel discovered that uranium salts emit radiation, and Marie Curie discovered radiation coming from the metallic element thorium and even stronger radiation from a mineral called *pitchblende.* She called the substances that gave off these rays radioactive. The Curies next undertook to discover exactly what chemical element produced the radioactivity. In July 1898, they announced the discovery of a new chemical element, which they named polonium after Marie Curie's homeland of Poland. The discovery of another element, radium, followed in December 1898. In 1903, Marie and Pierre Curie and Antoine-Henri Becquerel received the Nobel Prize in physics for their research on radioactivity.

The Curies' successful partnership was brought to an untimely end in 1906 when Pierre was hit by a horse and carriage while crossing the street. Marie Curie went on to study the chemistry and medical applications of radium, and in 1911 she was awarded the Nobel Prize in chemistry in recognition of her work in discovering radium and polonium and in isolating radium. In 1914, the University of Paris built the Institut du Radium (now the Institut Curie) to provide laboratory space for research on radioactive materials, and Marie Curie contributed to the advancement of scientific work at the institute by directing that the Nobel Prize money and other financial rewards be used to finance further research. The Curies refused to patent any of their discoveries as they wanted others to benefit from what they had learned.

Marie and Pierre Curie Museum, Warsaw *(Museum of Warsaw)*

During World War I (1914–18), Marie Curie played an active role in the use of radiation for diagnostic purposes. She helped equip ambulances with X-ray equipment and drove to the front lines herself. The International Red Cross made her head of its radiological service. She and her colleagues at the Institut du Radium also held courses for medical orderlies and doctors, teaching them how to use the new technique.

Marie's health eventually suffered from her work. The dangers of radioactivity were unknown at that time, and the Curies had never worn protective clothing or taken any measures to guard against exposure to radiation. Marie encountered vision problems from the radiation exposure, and she underwent several cataract operations, which was not an easy procedure at that time. She died of leukemia, almost certainly caused by radiation, on July 4, 1934. A few months before Marie's death, her daughter and son-in-law, the Joliot-Curies, had announced the discovery of artificial radioactivity.

(continues on page 53)

THE APPLICATION OF RADIOLOGY

In the century since Röntgen's discovery and Marie and Pierre Curie's work in the field, electromagnetic radiation has been put to medical use in several ways: for diagnostic radiology, to treat many kinds of cancer, and interventionally, to optimize treatment without surgery.

DIAGNOSTIC RADIOLOGY

Diagnostic radiology permits physicians to obtain both static (still) and dynamic (moving) images of body tissues and functions to study both normal anatomy and physiology and abnormalities caused by disease or injury. The process involves passing a localized beam of radiation through the part of the body being examined to produce an X-ray, which can take several forms. It can be a plain image such as the common chest X-ray; a mammogram (an X-ray image of the female breast used to scan for cancerous tumors); a tomograph (an image that reveals depth within a structure by using a series of X-rays); or a computed *tomography* (CT or CAT) scan, a computer analysis of a cross-sectional image of the body.

Some body parts (certain organ systems and muscular and skeletal structures) cannot be viewed using normal diagnostic radiology. However, physicians learned that if patients drink, inhale, or are injected with substances called contrast media, radiation can then be used to reveal these systems. Contrast media can be used to study the upper gastrointestinal tract, or a contrast substance can be injected into an artery, vein, or lymph vessel in order to produce an angiogram so that doctors can obtain more information about a patient's bodily functions. To capture these systems in action, radiologists can use fluoroscopy to obtain dynamic images of the intestinal tract or the flow of contrast material through blood vessels or the spinal canal. Fluoroscopy can either be analyzed while the test is being conducted or the images can be recorded for later study.

Positron emission tomography (PET) scans are used to diagnose brain tumors and strokes by injecting the patient with glucose with radioactive tracers. As the body metabolizes the glucose, the PET scan monitors the radioactive particles emitted by the tracers in the glucose. Since the 1970s, magnetic resonance imaging (MRI) is increasingly being used to produce computer-processed views of soft tissue, such as arteries, nerves, tendons, and some tumors, and ultrasound uses high-frequency sound waves, which are reflected by tissue in the body.

THERAPEUTIC RADIOLOGY

Therapeutic radiology uses ionizing radiation in the treatment of cancer. Normal tissues have a greater ability to recover from the effects of radiation than tumors and tumor cells. A radiation dose sufficient to destroy cancerous cells only temporarily injures adjacent normal cells. (Certain cancers are resistant to radiology, and then radiation is not part of the therapy.)

Radiation therapy is commonly employed either before or shortly after surgical removal of certain tumors to destroy tumor cells that could (or may already have) spread beyond the surgical margins. Radiation therapy may be used alone as the treatment of choice in most cancers of the skin, in certain stages of cancers involving the cervix, uterus, breast, and prostate gland, and in some types of leukemia and lymphoma, particularly Hodgkin's disease. Radiation is also used in conjunction with cancer-treatment drugs.

Scientists are exploring therapies for brain-controlled movement disorders such as tremors, epilepsy, and Parkinson's disease that would use targeted radiation.

(continues)

(continued)

INTERVENTIONAL RADIOLOGY

Interventional radiology uses radiologic imaging to guide catheters (hollow, flexible tubes), balloons, filters, and other tiny instruments through the body's blood vessels and other organs in order to bring about a solution without surgery. These types of interventional uses of radiology include balloon angioplasty, the use of a balloon to open blocked or narrowed arteries; chemo-embolization, the delivery of anti-cancer drugs directly to a tumor; fallopian tube catheterization, which opens blocked fallopian tubes, a common cause of infertility in women; and thrombolysis, which dissolves blood clots.

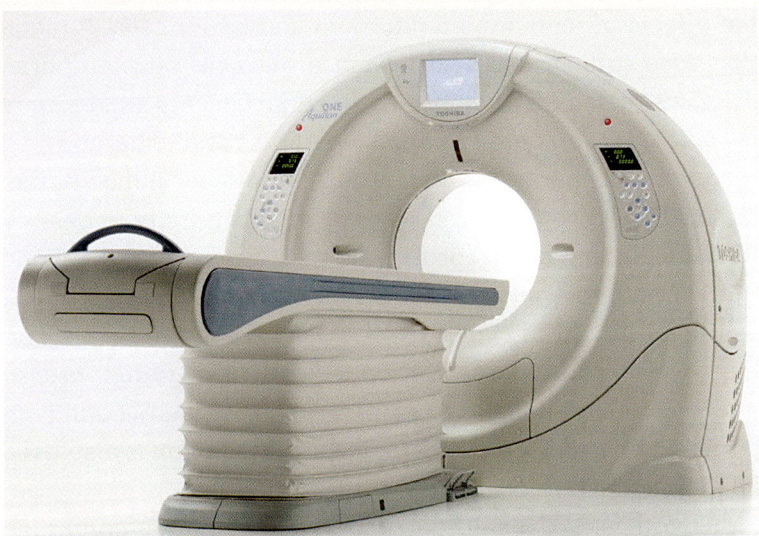

The Aquilion 320 provides a three-dimensional view of the intended organs in much less time than was previously required, thus reducing a patient's radiation exposure time. *(Johns Hopkins Medical Center)*

(continued from page 49)

Curie was one of the most famous women of her time. Though she resented the time fame took away from her work, she was able to use her good name to promote the medical uses of radium by facilitating the foundation of radium therapy institutes in France, Poland, and the United States.

THE DISCOVERY OF VIRUSES

After Louis Pasteur and Robert Koch identified bacteria as the cause of many diseases, most scientists became increasingly certain that it was just a matter of time until they unlocked the causes of every illness. Scientists identified a wide range of organisms, one by one, including those responsible for cholera, diphtheria, bubonic plague, and malaria, among others. However, there were still some diseases that left scientists mystified. Pasteur himself puzzled over one—rabies. Though he successfully created a vaccine to prevent rabies, Pasteur was never able to isolate or culture the agent that caused it. He speculated that whatever the entity was seemed to be too small for him to find.

The work of two men on behalf of the tobacco industry laid the groundwork for the field of virology. The process started in 1890 when a Russian graduate student in botany, Dmitri Ivanovsky (1864–1920), undertook a study of diseased tobacco plants to help Russian farmers reduce the damage occurring to their tobacco crops. A German botanist Adolf Mayer had identified that juice from sick plants could transmit illness to healthy plants, so Ivanovsky set out to find the causative microbe. In this early era of bacteriology, scientists had discovered that a filter could be used to trap the microbes they were investigating, and Ivanovsky selected an extraordinarily fine filter, known as a Chamberland candle, to use. Ivanovsky soon saw that whatever caused the illness was still too small to be stopped by his filter. He noted that the illness continued to spread from plant to plant, and he was unable to see the microbe, to catch it in a super-fine filter, or to grow it in a culture—all of which would have been possible had the agent been similar in size to bacteria. In 1892, he presented his findings to the St. Petersburg Academy of Science, concluding that the cause of the disease was either a toxin produced by the microbe or perhaps some type of minute microbial spore.

Farmers were still looking for help with the tobacco crops, and a Dutch botanist Martinus Beijerinck (1851–1931) began work on the problem. His experiments were similar to Ivanovsky's, but he added the element of heat. He heated the sap drawn from sick plants to 194°F (90°C) and found that the sap no longer caused illness. If the agent had been bacteria it would have survived the heating. He also noted that the alcohol and disinfectant that would have killed bacteria had no effect on the disease-carrying ability of the sap—therefore, it had to be a different causative agent. He termed it *contagium vivum fluidum* ("soluble living germ"). From Beijerinck's reference comes the term *virus*.

What are now known as viruses remained unseen, but scientists acknowledged that there was "something" there that was smaller than bacteria that was causing illnesses. Scientists also put forward the theory that only plants and animals were susceptible to these minute agents. In 1898, two German scientists were studying foot-and-mouth disease in hoofed animals, and they determined that this disease, too, was caused by this "soluble living germ," but they continued to believe that this type of germ did not invade the human species.

Viruses that infect bacteria (*bacteriophages*) were identified in the 1910s, and this permitted scientists to begin observing viruses' effects and how they multiplied. Finally, in 1931, with the aid of the newly invented electron microscope, scientists finally saw what a virus looked like, and the tobacco mosaic virus was one of the first to be studied in detail. Over time, scientists came to understand that viruses are subcellular organisms, which means they are smaller than most cells, including human cells and bacteria. (Bacteria can be measured in micrometers, viruses are described in nanometers.) Scientists were surprised to discover that most viruses consist of minute particles, not just fluid as had been suspected. Unlike bacteria, viruses are not living cells; they are tiny packets of genetic material that are parasitic. They must infect a host cell in order to reproduce and to manufacture substances for their own life cycle. Outside a host, some viruses can survive in a dormant state for quite some time and reactivate when absorbed into a new host. (Viruses have long been thought to be the smallest infectious agent, but recently that position has been taken over by two smaller pathogens known as *prions* and *viroids*.)

VACCINES PRECEDE EFFECTIVE TREATMENT

Physicians had some success in finding ways to prevent the spread of some of these illnesses. In 1798, Edward Jenner created the first vaccine against smallpox. Rabies was another disease for which a vaccine proved effective. Despite these gains, there was still a lot to learn, starting with a new discovery of how an illness could travel from person to person. In 1901, Walter Reed (1851–1902) and the Reed Commission, a team of investigators, confirmed the *vector* theory first conceived by Dr. Carlos Finlay. Finlay suggested that yellow fever was actually a highly contagious illness, but it did not pass from person to person unaided. The virus traveled from sick people to healthy people via a vector (carrier), in this case, the mosquito. Since scientists could not identify a bacteria connected with the disease, this was the first proof that a virus caused illness in humans.

As technology has improved, scientists have been able to undertake detailed identification and characterization of viruses. They now know that viruses enter the body in many ways—for example, they can be ingested, inhaled, or passed person to person via a vector. Physicians were fortunate that they gained a weapon against some viruses through the creation of vaccines, as viruses are hard to cure once they invade the body. They take up residence inside the body's own cells, making it difficult for the immune system to attack them. It was not until the 1970s that there were drugs available to the general public that were helpful in curing viruses.

Scientists are still baffled by many viruses. Many types of viruses can mutate quickly, so the search for a cure is often very fluid as the virus changes in order to survive. SARS (severe acute respiratory syndrome) and HIV (human immunodeficiency virus) are two of the more troubling viruses that still await answers today.

NEW METHODS OF TREATMENT

Emil Behring (1854–1917) was the pioneering doctor who introduced the first serum therapies. The use of serum therapies involved immunizing animals against specific diseases and then extracting blood serum from those animals to inject into people with the disease. Behring primarily focused on diphtheria and tetanus, two diseases that were untreatable at the end of the 19th century. In the late 19th century, statistics for Germany alone reveal that 50,000 children died annually from the disease.

Behring started his medical career in the military, but he became interested in research when he had the opportunity to work under the renowned Robert Koch at the Institute for Infectious Diseases. Behring began experimenting using a vaccinelike method to create a cure. Working with rats, guinea pigs, and rabbits, Behring immunized the animals to create a weakened form of the infectious agent that caused diphtheria. The researchers then extracted the blood serum from these newly immunized animals and injected it into nonimmunized animals that were infected with diphtheria. They discovered that the sick animals could be cured with the serum. Behring and a colleague were credited with developing the first effective therapeutic serum against diphtheria. Behring and another colleague Shibasaburo Kitasato went on to develop a serum against tetanus. Neither, however, had yet been tested on people. Behring's early tests using the serum on people were unsuccessful as he had not yet mastered how to create the proper concentration of the *antitoxins*—strong enough to fight the disease but not so strong to add to the infection.

Working separately, bacteriologist Paul Ehrlich (see chapter 4) developed a standardized way to determine the quality and quantity of antitoxins to use in the serum. By 1894, a pharmaceutical company was producing the serum.

Behring went on to explore whether he could create a therapy to use against tuberculosis. He soon realized this was not feasible, and he applied his knowledge to creating vaccines to use against diphtheria and tetanus.

CONCLUSION

The progress in medicine during this era—the late 1800s and early 1900s—was truly remarkable, but different segments of the population would give very different answers if they were asked to identify the most significant advance during the time period. Certainly, some scientists would point to the discovery of radiation that is now used diagnostically, therapeutically, and interventionally; others would highlight the discovery of viruses. But anesthesia would undoubtedly be cited by patients as the most significant progress of the era. At last, patients could receive surgical treatment without undergoing the searing pain of the process.

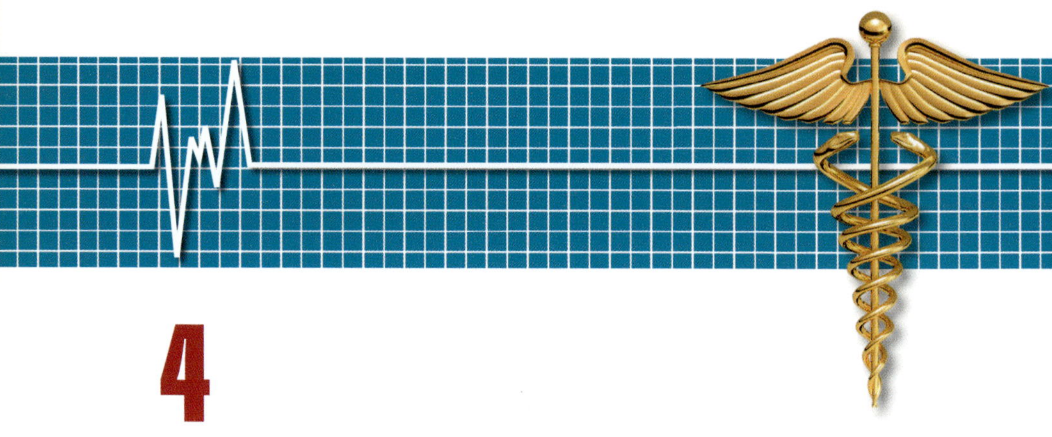

4

Advances in Medications

When scientists determined that bacteria was often the cause of illness, this marked a major step forward because now scientists and physicians had a new lens through which to consider their ideas for treatment. They realized that medications that made a patient feel better were valuable but that they did not necessarily bring an end to an illness. For that they needed something that could actually kill bacteria. The search for these medications, now known as *antibiotics,* began in the late 1800s.

The early sulfonamide drugs (sulfa drugs) helped manage infections until antibiotics were discovered. The sulfa drugs inhibit bacterial growth and activity by interfering with the metabolic processes in some types of bacteria. The development of these drugs showed that there were medications that could fight several strains of infection. In the 1930s, when the first sulfa drug, Protonsil, was shown to be effective, a sulfa drug craze ensued. Pharmaceutical companies were particularly excited when they realized that the prime active ingredient was not patent protected, meaning that any company could make it. The sulfa drugs saved thousands of lives in World War II by providing something that could help fight infection in wounded soldiers.

But medicine was truly transformed by the accidental discovery of penicillin, which marked the start of modern antibiotics. Initially penicillin and the antibiotics that followed were greeted as miracle drugs that could wipe out serious illness, and, of course, they were. However, today physicians are much more cautious. They now have witnessed that a number of patients develop allergies to the medications from overuse, and, more frighteningly, bacteria have proven capable of evolving and developing into what are unscientifically referred to as superbugs, which are not controlled or eliminated by the current arsenal of antibiotics.

Three other drugs are worthy of highlighting in this chapter. The first, aspirin, preceded the development of penicillin, and physicians still marvel that this very simple medication is highly effective; scientists today continue to explore ways it can be used. This chapter introduces Salversan, a drug that was developed by the scientist Paul Ehrlich. Salversan was the first effective drug against syphilis, and it is particularly significant because it is considered the first of the *chemotherapy* drugs, which are primarily used to fight cancer.

The final medication that is described in this chapter is one that also changed history—the birth control pill. For the first time, women were given an increased ability to choose their own fate by managing their childbearing. This has had an enormous effect on developed countries, and the manner in which the medication was created required the confluence of both science and social opinion.

ASPIRIN: SIMPLE AND EFFECTIVE

As long ago as Hippocrates' time, people knew that there were ingredients in nature that could be used to ease aches and pains. In the Egyptian Ebers papyrus that dates back 3,500 years ago, it was noted that the juice from the bark of the willow tree could ease discomfort. This same recommendation appeared in the 1700s, when the scientist Reverend Edmund Stone wrote of using willow bark to help with fevers and "agues." As scientists began to

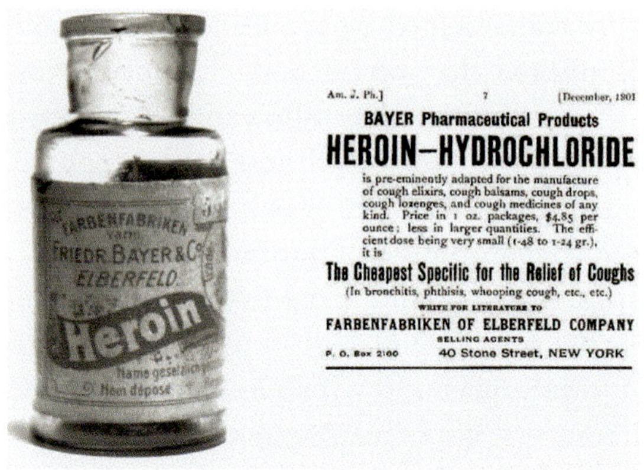

In addition to aspirin, one of Bayer's early products was heroin—as the advertisement indicates, it was highly regarded for relieving coughs.

explore the chemical makeup of this cure-all, the useful chemical in the willow bark came to be known as salicin (the name is derived from *salix*, Latin for "willow").

The next challenge was how to make the chemical into something that could be proffered as a medication. An Italian chemist by the name of Piria began working to turn salicin into an ingredient that could be taken medicinally. He found that he could create salicylic acid, and it worked very well to reduce pain and swelling. The only hindrance was that this new medicine often caused patients to have upset stomachs. For some people, the stomach pain was only a mild ache, but other people experienced bleeding from the digestive tract, forcing them to stop treatment. In 1853, a French chemist named Charles-Frédéric Gerhardt tried to improve on the sodium salicylate concoction by combining it with acetyl chloride. He produced a new compound that was less irritating to the stomach, and he published a paper on the topic. However, he soon abandoned the project because the process of making the new preparation was time-consuming and tedious. (Some experts think Gerhardt deserves more credit than he gets in the creation of aspirin.)

Felix Hoffmann, a German chemist employed by Friedrich Bayer & Company, is the person credited with coming up with a way for salicylic acid to be taken with fewer negative side effects. Hoffmann's father suffered from terrible joint pain, and Hoffmann saw firsthand how uncomfortable his father was when taking salicylic acid for his pain. He began looking for a way to reduce the stomach distress. Hoffmann assumed that the damage was occurring because the drug was an acid, so he worked to find something that would mask the acidic aspect of the drug without harming its ability to help with pain and swelling. He eventually discovered that he could cover the acidic parts by converting it to acetylsalicylic acid (ASA), which seemed to work well for patients without damaging the stomach.

Though today the names "Bayer" and "aspirin" are deeply interconnected and the worldwide market for this over-the-counter drug is huge, Bayer did not respond enthusiastically to Felix Hoffmann's discovery after he revealed what he was working on. It turned out that Hoffmann and his drug became entangled in a company turf war. Hoffmann worked in the research department under a fellow named Arthur Eichengrün; Eichengrün and Hoffmann had contracts with Bayer by which they would receive a royalty on any patentable product they invented, and Eichen grün knew that Hoffmann's discovery held that potential. They encountered stalling and resistance from Heinrich Dreser, who was in charge of Bayer's pharmacological testing and standardization department. Dreser had an agreement with Bayer by which he would receive a royalty on any product he introduced. If Dreser could "introduce" the project, and Hoffmann and Eichengrün never patented it, they were out of the picture. When the drug finally entered the marketplace under the Bayer name, Dreser was the financial beneficiary. This was the way it eventually worked out. As for calling it aspirin, the people at Bayer came up with the term: "A" was for acetyl chloride, and "spir" was for the spiraea ulmaria plant (the plant from which the salicylic acid was derived), and "in" was added as a common ending for medications.

While aspirin was the perfect drug for wide introduction since it had many applications and could help with many types of pain, the Bayer Company was also the perfect company to be the first major manufacturer of it. Bayer had been started in 1863 as a dye-manufacturing plant, but as the dye industry began to fade during the 1880s, the company decided to switch from dyes to pharmaceuticals. However, from their past as a dye company, they were accustomed to a business model that involved mass production and marketing on a scale far beyond what drug companies had done in the past. Bayer had their own sales representatives, they took out trade ads, and they assumed that their work should be patented just as it would be in the dye industry. These practices were quickly incorporated by the pharmaceutical industry. Drugs became commodities that needed to be standardized and were manufactured in an industrial setting for profit.

Early advertising for aspirin stressed safety, particularly heart safety. Today scientists know that aspirin can actually be beneficial for heart health but early on, it was actually a top concern. During the 1920s, when only one brand of aspirin existed, Bayer advertisements promised: "Does Not Affect the Heart."

Bayer was extremely successful with its new drug, but its exclusivity was to end after World War I. Bayer was a German company, and at the end of World War I part of Germany's agreement in

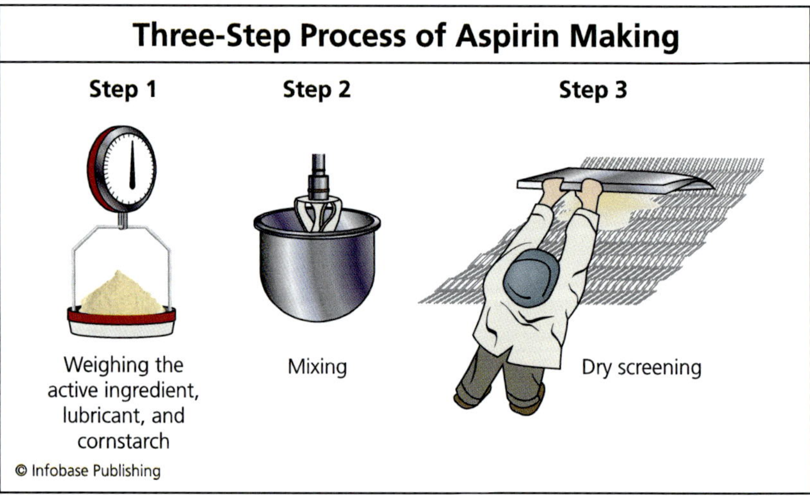

Cornstarch and water are added to the active ingredient, acetylsalicylic acid, to create a pill form of the medication. The cornstarch serves as binder and filler, and a portion of the lubricant, ranging from vegetable oil to a water-soluble acid, keep the mixture from sticking to the machinery. The mixture is dried and then formed into tablets.

concluding the war involved reparations by specific companies. Bayer had trademarks on aspirin as well as heroin (at this time, heroin was used medicinally as a very powerful cough suppressant and pain reliever) and had to give up both trademarks to France, England, and Russia. This agreement was actually part of the Treaty of Versailles.

The Evolution of the Use of Aspirin

During the 1880s and 1890s, physicians were very concerned about the negative effects of fevers; a high body temperature was considered very worrisome. Physicians reasoned that fever indicated that patients' tissues were burning up. Aspirin successfully brought down fevers, and, because it was so effective, this was one of the first uses for which it was touted. However, as medical study advanced, physicians began to note that even when aspirin lowered a patient's fever, it did not necessarily change the course of the disease. Over time, they began to realize that if the aspirin did not alter the outcome of the disease, there was probably less need to focus on bringing down the fever.

Aspirin was still viewed as a helpful medicine, and the focus returned to the benefit of taking it to relieve pain, a use for which it is still recommended. Now aspirin is manufactured by numerous companies in various forms and is one of the most widely used drugs in the world. Though aspirin is still a vital tool in the fight against pain and is often recommended in cases where physicians want a patient's fever to be brought down, today the majority of people who take aspirin do so for heart health. Americans alone take at least 80 million aspirin tablets a day.

The remaining side effects of aspirin that persist have to do with stomach aches or actual harm to the stomach lining (see the following sidebar "How Aspirin Works" on page 64). Scientists have introduced drugs such as acetaminophen and ibuprofen and naproxen. (The latter two are known as nonsteroidal antiinflammatory drugs, referred to as NSAIDS.) These are often used for people who should not take aspirin, and while they are helpful with pain relief, none have the same protective effect in guarding against heart disease.

As scientists contemplate the future of aspirin, they continue to marvel at its versatility. They are finding that it can help with high blood pressure, guard against strokes, be mixed into a salve to relieve itching, calm the irritation from a sunburn, reduce the risk of heart attack, soften calluses, shorten the length of time a cold

HOW ASPIRIN WORKS

When physicians began recommending aspirin (in the form of salicylic acid) for pain or to bring down fevers, they had no idea how it actually worked. Doctors knew it was not site specific; if a patient's head ached, the aspirin still traveled through the whole body, and yet the only noticeable effect on the person had to do with the reduction of the pain in the person's head. It was not until 1982 that a scientist grasped how aspirin worked, and the British pharmacologist Sir John Vane won the Nobel Prize in physiology or medicine for his discovery.

The first step in understanding how aspirin decreased pain had to do with learning more about how people register their sense of feeling. Over time, scientists came to realize that no one feels anything until a message is sent to the brain via a nerve that is designated to take in various information—whether it is the feeling of warmth, cold, pain, something rough against the skin, or the sensation of a blow to the body.

The next thing Sir John Vane came to understand was that when the body is hurt, the damaged tissue releases message-sending chemicals, some of which are prostaglandins. Based on the strength of the chemical signal, the brain tells the body how to respond, which can vary from telling the hand to move away from a hot pan or to simply register that the person has a headache and the person may rub his forehead as a result.

Vane realized that prostaglandins were key, but he had more unraveling to do because he knew the body pro-

sore lasts, and may decrease the odds of a person getting Alzheimer's by reducing inflammation in the brain.

Aspirin may also prove helpful with cancer. Researchers at Newcastle University in England explored a biological process that makes blood vessels grow, and they have observed that aspirin

duces two types of prostaglandins. Some prostaglandins are released by damaged tissue and create swelling and send pain messages; others help to provide a protective lining for the stomach and small intestine. The prostaglandins that are created from a painful incident are created using an enzyme called cyclooxygenase 2 (COX-2). Eventually, Vane learned that aspirin has the ability to stick to COX-2, reducing the creation of prostaglandins. Aspirin does not correct or cure a headache or arthritis or totally reduce pain, but it does reduce the number of pain signals traveling through the nerves to the brain.

When a patient takes an aspirin, it dissolves in the stomach or the small intestine and is absorbed into the bloodstream. Although it travels throughout the body, it only works where there are prostaglandins being made, and this includes both the area where there is pain as well as the stomach. Because aspirin is nonselective, it prevents prostaglandins from being made both in the stomach and in the area experiencing discomfort. When aspirin is taken too often or in too great a quantity, it can cause thinning of the stomach lining, and this is why aspirin can cause stomach pain or bleeding. People have always suspected that aspirin hurt the stomach because it was an acid, but stomach problems are actually the result of the reduction of prostaglandins needed in the lining of the stomach.

seems to block the formation of blood vessels that feed the growth of cancer. This needs further study, but it is one more amazing application for a very simple, very old drug.

THE DISCOVERY OF PENICILLIN CHANGES MEDICINE

An accidental discovery by the Scottish biologist and pharmacologist Alexander Fleming (1881–1955) changed the world of medicine. As Fleming identified the power of what he and Howard Florey and Ernst B. Chain would eventually develop into penicillin, physicians suddenly had a drug that could fight bacterial infections—a very important first that was to revolutionize medical care.

Alexander Fleming was born into a farming family in Scotland. His father died while Fleming was very young, so Fleming was greatly affected by the life paths of his brothers. One of them became a physician, and Fleming followed his path to medical school. He was particularly influenced by one of his professors, Sir Almroth Wright, a pioneer in vaccine therapy and immunology. Fleming became interested in the natural bacterial action of the blood and in antiseptics and, after serving as a captain of the army medical corps during World War I, he returned to St. Mary's Hospital in London to search for antibacterial substances that would not be toxic to animal tissues.

Fleming's war experiences had influenced him greatly; he saw many soldiers die from *septicemia* from infected wounds. The treatment of wounds at that time involved cleaning the wounds with strong antiseptics, which had an unfortunate result. In addition to killing bacteria on the surface of the wound, the antiseptics also weakened the soldier's immunological defenses. Fleming also saw that bacteria in deep wounds were not affected by the antiseptic agent. During World War I, Fleming had an article about this problem published in the respected medical journal the *Lancet,* but it did not stimulate discussion or change. Physicians on the battlefield did not change their practices; there was actually not much they could do because there were no good alternatives.

In 1928, Fleming noticed something that was to lead to a significant discovery. He was cleaning up his laboratory from some earlier experiments when he observed that mold had developed on a staphylococcus culture plate. Inexplicably, the mold had created a bacteria-free circle around itself—the staph infection culture that was growing elsewhere in the plate disappeared around the mold. Curious, Fleming grew the mold (*Penicillium notatum*) in a pure culture and found that it produced a substance that killed a number of disease-causing bacteria.

Sir Alexander Fleming (1881–1955) discovered penicillin in 1928, thus making one of the greatest contributions to medicine.

He was inspired to experiment further, and he found that a mold culture prevented growth of staphylococci, even when it was diluted. He continued his research and determined that the mold could kill other types of bacteria. In experiments with small animals, it seemed to have no ill effect. Fleming published his discovery in 1929 in the *British Journal of Experimental Pathology*. He noted that his discovery might have therapeutic value if it could be produced in quantity. Fleming continued working but found that cultivating it was quite difficult, and he was unsuccessful at isolating the antibiotic agent involved. As he experimented, he found the *penicillium* he was working with was slow to act, and he determined it would not last long enough to fight bacteria in the human body. Fleming also worried that it would be hard to produce in quantity.

Other research tasks interrupted Fleming's work on *penicillium,* and it was a full 10 years later that two Oxford University scientists Howard Florey and Ernst Chain took the investigation a step further and isolated the substance in the mold that killed the bacteria, now known as penicillin. Together, they worked out how to isolate and concentrate penicillin, and an associate Norman Heatley came up with a method of maintaining the active ingredient so that it could be delivered into the bodies of animals or humans.

As the laboratory work progressed, penicillin still had not faced an ultimate test of whether or not it could save lives by curing infections. The physician Charles Fletcher, who worked in a hospital near Oxford where Florey and Chain were based, heard of their work and approached them on behalf of one of his patients. Fletcher had a patient with a severe bacterial infection who was dying—nothing the physician tried had helped. Fletcher connected with Chain and Florey and asked for some penicillin to try with the patient. Florey and Chain quickly agreed, and, to the delight of the scientists and the physician, the patient's condition improved. Unfortunately, the importance of dosage and quantity were not understood. Though the patient got better in the beginning, the dosage was not enough to kill all the bacteria in the patient's body, and he eventually succumbed to the infection. Though the result of this case was disappointing, the experience had provided a great deal of information about penicillin. They knew it worked; they just needed more of it.

As World War II continued, the men knew that penicillin could be key to helping many of the wounded, but they needed a way to get it produced in quantity. England was overwhelmed by the war, which was draining industrial and government resources, so Florey made contact with friends in the United States who referred him to a lab in Peoria, Illinois, that was already doing some work with growing fungal cultures. On July 9, 1941, Howard Florey and Norman Heatley traveled to Peoria carrying a small quantity of penicillin. The scientists at the American lab already had created a way to speed fungal growth that involved submerging mold and several chemical ingredients in deep vats into which they pumped air. They also encouraged Florey to experiment with other sources of mold, and, as it happened, the sturdiest fastest-growing mold

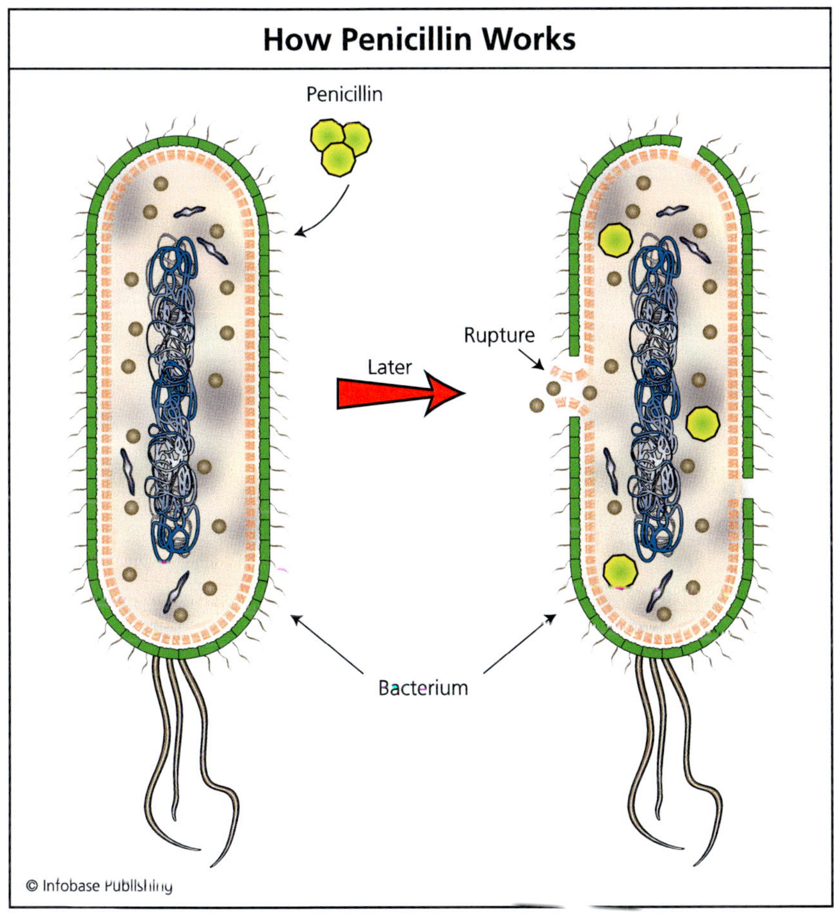

Penicillin kills bacteria by interfering with its ability to synthesize cell wall material. The bacterium lengthens in preparation for dividing, but eventually the weak cell wall ruptures, and the bacteria cannot multiply.

came from a strain of penicillin from a moldy cantaloupe purchased by Florey in a Peoria grocery store.

By November 26, 1941, Dr. Heatley and Andrew J. Moyer, one of the lab scientists familiar with mold, succeeded in greatly increasing the yield of penicillin that was possible, and, for the first time, it seemed viable that penicillin might one day be mass produced. In 1943, penicillin was tested in clinical trials, and it was shown to be the most effective antibacterial agent anyone had yet seen. The number of injured soldiers fighting for the Allies was very much on everyone's minds, and penicillin production was increased quickly. With increased production, the price dropped from an astronomical per dose price in 1940 to $20 per dose in July 1943 to $0.55 per dose by 1946.

Penicillin soon became known as the wonder drug, and in 1945 Fleming, Chain, and Florey were awarded the Nobel Prize in physiology or medicine, thereby ushering in the era of antibiotics. On May 25, 1948, Andrew J. Moyer was granted a patent for a method of mass production of penicillin.

THE CREATION OF OTHER ANTIBIOTICS

An antibiotic is a chemical substance produced by one organism that is destructive to another. The chemist Dorothy Crowfoot Hodgkin used X-rays to find the structural layouts of atoms and the overall molecular shape of more than 100 molecules including penicillin. Hodgkin's discovery of the molecular layout of penicillin contributed to scientists' ability to develop other antibiotics. Some of the early or more common ones include the following:

- **Tetracycline.** This is a large family of antibiotics that is used to treat a broad spectrum of illnesses; it was first observed in plants by the American physiologist Benjamin Minge Duggar. Lloyd Conover, a scientist working at Pfizer, patented the antibiotic in 1950, and it became a widely prescribed medication. Tetracycline sparked the development of many chemically altered antibiotics. Today, it is often prescribed to treat acne, but when it was first discovered, it played an important role in stamping out cholera in the developed world.

- **Nystatin.** This drug was patented in 1957. It is often used as prophylaxis in patients who are at risk for fungal infections, such as AIDS patients and patients receiving chemotherapy. Like many other antifungals and antibiotics, nystatin was developed from a bacteria. It was isolated from *Streptomyces noursei* found in a soil sample in 1950 by Elizabeth Lee Hazen and Rachel Fuller Brown, who were employed by the Division of Laboratories and Research of the New York State Department of Health. In 1954, Hazen and Brown named nystatin after the New York State Public Health Department. In addition to human ailments, the drug has been used to treat such problems as Dutch elm disease and to restore water-damaged artwork from the effects of mold. The two scientists donated the royalties from their invention, more than $13 million, to a nonprofit corporation for the advancement of academic scientific study.

- **Amoxicillin.** SmithKline Beecham patented Amoxicillin or amoxicillin/clavulanate potassium tablets in 1981 and first sold the antibiotic in 1998 under the trade names of Amoxicillin, Amoxil, and Trimox. Amoxicillin is a semi-*synthetic* antibiotic. It is a commonly prescribed drug because it is administered orally and is well absorbed and well tolerated by most people.

When antibiotics were first introduced, no one predicted that overuse of these medications would cause the emergence of stronger, antibiotic-resistant bacteria. The mutations in these infectious agents pose huge dangers for the future. Currently, government health groups worldwide are recommending that doctors curb the number of antibiotics prescribed and use them only when no other remedy will help.

THE SEARCH FOR A MAGIC BULLET

Paul Ehrlich (1854–1915) was a German chemist and bacteriologist who made progress in many fields. He worked with Emil Behring (see chapter 3) on developing an antitoxin to diphtheria and pioneered work in blood and histology that established hematology as a field. But his search for a magic bullet to find chemical substances that would specifically target pathogenic organisms launched the field of chemotherapy. In the process, Ehrlich developed a very important drug that was key to helping the many people who had syphilis, which until that time was incurable.

SUPERBUGS AND RESISTANCE TO ANTIBIOTICS

The widespread use of antibiotics to fight bacterial illness beginning in the 1940s revolutionized medical care and dramatically reduced illness and death from infectious diseases. Today, however, many bacterial infections in the United States and throughout the world are developing resistance to the most commonly prescribed antibiotic treatments.

The first microbes that appeared to be resistant were noted very early. In 1947, just a few years after drug companies began mass-producing penicillin, scientists saw that some microbes were resistant. The first to be identified was *Staphylococcus aureus,* a bacterium that is often a harmless passenger in the human body. However, it can cause illness, such as pneumonia or toxic shock syndrome, when it overgrows or produces a toxin. A succession of other microbes have followed.

Antibiotic-resistant infections are increasingly a problem for hospitals and nursing homes where they can spread from one patient to another, taking advantage of open wounds and suppressed or overtaxed immune systems. In its October 2007 issue, the *Journal of the American Medical Association* estimated that 94,360 U.S. patients developed an invasive infection from antibiotic-resistant *Staphylococcus* (MRSA) in 2005 and that nearly one of every five, or 18,650 of them, died.

Sometimes called superbugs, these infections cannot be successfully treated with commonly prescribed antibiotics

The *spirochete* that causes syphilis was discovered by two researchers in Berlin. Ehrlich decided to look for a drug that would be effective against this particular bacterium and began exploring some arsenic-based drugs. Though arsenic was poisonous, Ehrlich

and often involve longer illnesses, extended hospital stays, severe side effects from last resort drugs, and lead to higher treatment costs. Increasing numbers of U.S. soldiers returning from Afghanistan and Iraq have been plagued with highly resistant infections. In addition, several cases of antibiotic-resistant infections have recently occurred in the general population. The elderly, the immuno-compromised, and the very young are most vulnerable, and some cases are virulent enough to cause fatalities in young children.

A major factor in the emergence of antibiotic-resistant bacteria is the overuse and misuse of antibiotics. Most illnesses are caused by two kinds of germs: bacteria and viruses. Antibiotics are effective against infections caused by bacteria, like a strep throat. Antibiotics do not work against viruses, like the common cold, the flu, and the majority of sore throats and runny noses, even though sometimes physicians will "take a chance it will help" and prescribe them.

"This overuse threatens the effectiveness of these precious drugs," says Dr. Cindy Friedman, a medical director at the Centers for Disease Control. "Doctors and patients are both part of the problem. Studies show that if a doctor believes a patient wants an antibiotic, he or she is much more likely to prescribe one, even if the patient doesn't really need one."

hoped he could find a magic bullet that was fatal to bacteria but not fatal for humans. When a Japanese bacteriologist Sahachiro Hata (1873–1938) came to work at Ehrlich's institute, he brought some experience that intrigued Ehrlich. Hata had succeeded in infecting rabbits with syphilis, so Ehrlich and Hata began to review some of the arsenical drugs that had been set aside as ineffective. When Hata tried 606 (the 606th compound out of 900 tried), it cured the rabbits. Hata and Ehrlich then ran experiment after experiment until they were sure it worked. When Ehrlich announced its success, he called it Salvarsan.

Like many other drugs, the medication was not accepted immediately, but eventually Salvarsan or Neosalvarsan (the more easily manufactured and better tolerated version) were accepted for the treatment of human syphilis. At last, the medical world had a way to treat the devastating illness of syphilis, and Ehrlich laid the groundwork for future forms of chemotherapy.

THE ORAL CONTRACEPTIVE PILL

Few medications have created the stir that the pill has. It changed the lives of women by permitting them to control the number of children they have—something that was impossible in earlier times. (This form of birth control is viewed negatively by many religious groups.) Many of the issues surrounding birth control have become hot button issues for modern society. While the work of several scientists made the creation of an oral contraceptive possible, the work of a single woman must be acknowledged first— without Margaret Sanger the social conditions for the scientific advancements would not have been in place.

Margaret Sanger (1879–1966) was born into poverty in Corning, New York, and was one of 11 children in a Catholic family. Her mother had been pregnant 18 times, and seven of the pregnancies had ended in miscarriages. She died at the age of 50 from tuberculosis. Margaret was 19 at the time and was said to feel that her mother's numerous pregnancies had been a major factor in her ill health and subsequent death. Margaret also noted class differences in Corning. She observed that the wealthy families had fewer children and seemed to have fewer problems, while poorer families had more offspring and also encountered other difficult family issues including unemployment, drunkenness, and fighting.

With help from her sisters, Sanger worked her way through nursing school and married a young architect. When Sanger first began her nursing career, she frequently was asked by poor women for the secret that rich women, who had fewer children, seemed to know. By law, only doctors could talk about birth control; nurses were not to comment. In addition, the only methods available to doctors for women who wanted to limit their pregnancies were condoms and diaphragms—what are known as barrier methods. If a woman's doctor did not tell her about these methods, it was not easy to get the information. Federal and state laws prohibited birth control information to exist in print. Both married women and single women frequently were desperate to avoid childbirth, and when they became pregnant and felt they could not cope with a pregnancy, they went to back-alley abortionists. Some died during the abortion; others died afterward from infection. Those who survived often had difficulty having children in the future.

Sanger and her husband started a family but rejected the normal suburban lifestyle. Margaret and her husband spent time in Greenwich Village and were friends with social activists, free thinkers, labor activists, and women's rights leaders. Sanger happened to attend a lecture by the famous psychiatrist Sigmund Freud (1856–1939), and, for her, his comments on sexuality and sexualization of women provided inspiration to fight against female repression. To Sanger, this social subjugation began with the lack of control women had over their bodies. In 1914, she published a newspaper *The Woman Rebel* that offered a platform for her views and in the June issue she used the term *birth control* for the first time.

Since there were laws that banned any written material about ways to prevent conception, the government indicted Sanger on an indecency charge and pronounced her publication obscene. If found guilty, Sanger could have been sentenced to 45 years in prison, so she fled the country, leaving her husband to take care of their children. William Sanger was soon jailed for handing out copies of his wife's book *Family Limitation,* so Margaret decided her only option was to come back to face trial. By this time, there had been an outpouring of support for Sanger, and the government, receiving a negative response to what they were doing, dropped the case.

Sanger's next step was to open a birth control clinic, the first ever in the United States. She faced condemnation by the Catholic Church, and sentiment against the presence of a birth control clinic was strong enough that police raided the facility only 19 days after it opened. Sanger and one of her sisters who worked with her were tried and sentenced to 30 days in jail for distributing contraceptive information. Both women refused to eat while incarcerated and had to be force-fed for the length of their sentence. Shortly after their release, there was a glimmer of hope. The New York State Court of Appeals broadened the law's definition of disease to incorporate the risks of pregnancy and thus began the process of legalizing contraception.

Sanger continued her social activism, starting clinics and creating the Planned Parenthood Federation. All the attention heightened interest from a number of people. Katherine Dexter McCormick (1875–1967), a wealthy woman whose husband was part of the family who ran a major agricultural machinery manufacturing company, was one of the first to step forward to offer help. McCormick's husband developed schizophrenia, and McCormick feared the disease could be passed genetically, so she did not want children. This gave her a vested interest in the cause, and she offered to fund research for the search for an oral form of birth control that women could take regularly and easily.

The work of several scientists contributed to the ultimate development of an oral contraceptive. A scientist named Gregory Pincus (1903–67) had done research in genetics and embryology. He had encouraged cell division in rabbit embryos, a precursor of cloning. This was not well received by Harvard where he was teaching, so he was denied tenure. He and a scientist friend started a company, the Worcester Foundation for Experimental Biology. Sanger felt Pincus's expansive thinking made him right to approach about looking for a safe oral contraceptive. Pincus was familiar with research that had identified progesterone, and researchers at the University of Rochester had shown that progesterone prohibited ovulation. This was further extended in 1937 by researchers at the University of Pennsylvania who successfully used progesterone to block ovulation in rabbits.

Pincus knew the next thing he needed was affordable synthetic hormones, and he turned to Russell Marker, an organic chemist at the University of Pennsylvania who developed a method of creating hormones from natural substances. Marker was soon replaced by the chemist Carl Djerassi who extended the method and perfected an oral form of progesterone that was effective at halting ovulation.

Next, Pincus added a physician to his team, a Catholic doctor named John Rock (1890–1984) who specialized in fertility problems. Rock had angered the Catholic Church by supporting the right for medical doctors to talk about contraception, and he wrote a book called *Voluntary Parenthood*. Rock was an able and well-qualified person to work on the product. Pincus and Rock wanted to establish some drug trials to see how it worked in women. A network of health clinics in Puerto Rico permitted the trials to be conducted and asked for volunteers from their patient base. Women came eagerly to get what became known as the Pill. The statistics gathered from these trials were promising. Pincus and Rock noted that it seemed to be 30 times more effective than other forms of birth control, but 17 percent of women suffered side effects (some of which were quite debilitating, such as daily vomiting). The possibilities of these side effects did not deter women, the scientists, or even the Food and Drug Administration, the governmental body that needed to approve the drug before it could be marketed. When Pincus's group approached the FDA, the approval actually went through quite easily. One of the main ingredients, Enovid, had been approved in 1957 for the treatment of severe menstrual bleeding, and the Pill itself was approved on May 11, 1960.

Today there are many versions of the birth control pill. The efficacy remains quite good and scientists have found ways to reduce the side effects by creating a medication that uses the lowest hormone dosage possible. While there is still a degree of risk to those who take the oral contraceptive, including a slight risk of blood clots, strokes, and cancer in some women, on the whole it is a well-regarded medicine.

The availability of a birth control pill has changed the lives of women in developed countries. Women in poor countries still face the same problems women in the United States faced 100 years ago. They have difficulty preventing pregnancy, and abortions contain a big element of risk because they are still often done by amateurs.

Despite the availability of an oral contraceptive for almost 50 years, the issues revolving around birth control are far from settled. While most women would agree that the oral contraceptive offers a great improvement over the inconvenience of the diaphragm (since the condom also prevents against disease, it should be used even if other forms of birth control are in place), the fact that a woman must be on a medication for 365 days a year in order to prevent pregnancy has its drawbacks. It also places full responsibility for birth control on the female.

As the field progresses, the morning-after pill (RU-486) is now available by prescription. This drug, known as Mifeprex, causes a potential fetus to abort. As of 2006, the FDA approved Plan B as a nonprescription, behind-the-counter medicine that could be sold to adults (including men) as an after-the-fact, emergency contraceptive. In 2009, the courts ruled that the FDA should make Plan B available to girls 17 and older without a prescription. (This lowers the age by one year for those who can buy it without a doctor's prescription.) Scientists are at work on a male contraceptive pill that interferes with sperm production but shows no long-term effects once the medication is stopped. It is anticipated that it will be available within the next five years.

CONCLUSION

As the cause of illnesses became better understood, scientists finally could make notable inroads in creating pharmacological treatments. Antibacterial medications like penicillin and other antibiotics meant that more patients could survive serious infections, and the search for a "magic bullet" not only gave the medical world a way to cure syphilis but Salvarsan was the first of the medications now known as a chemotherapy treatment, a class of drugs that creates the main arsenal against cancer. The development of the oral contraceptive pill made a major difference in women's lives by giving them an opportunity to control child-bearing, and there are doubtless more gains in store in this field as efforts are made to create a male contraceptive pill. Not to be forgotten is the basic wonder drug of the chapter, aspirin, an inexpensive nonprescription medication for which scientists continue to find new uses.

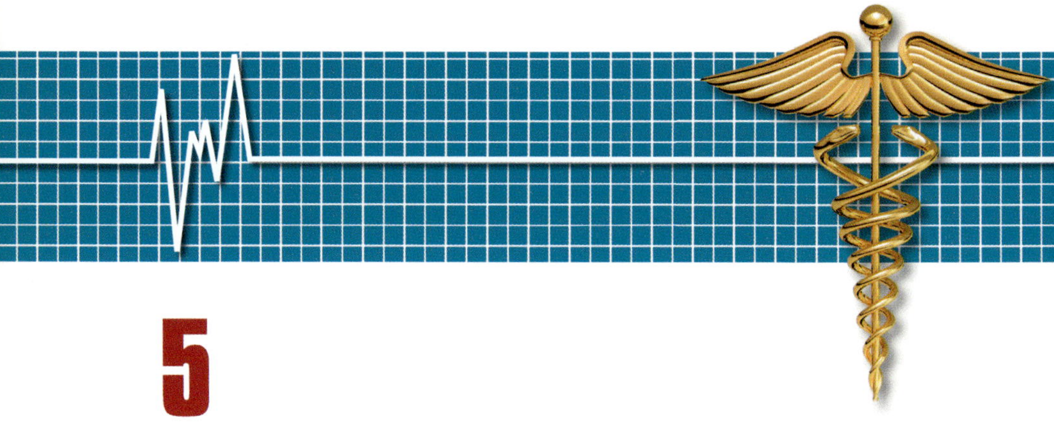

5

An Answer to Polio and Other Changes in Medicine

The world of medicine changed enormously during the 20th century. Scientists and physicians solved major health threats like polio, but there were general weaknesses in the overall approach to medicine. Leaders in the field began to see that without a decision-making system, physicians were only a little better off at making diagnoses than the physicians of 50 years ago.

Infantile paralysis, or *poliomyelitis,* has been around for a long time, but it did not make headlines until the 20th century. The most famous polio victim was Franklin Delano Roosevelt (1882–1945), who contracted polio in 1921 but went on to be elected to an unprecedented four terms as president of the United States (1933–45). Roosevelt was outspoken about the need to raise money to prevent or cure the disease, but he went to great lengths to hide from the public the true nature of his illness. He was never photographed in his wheelchair, and when he arrived at public events, he usually had a son and an aide to help him as he appeared to walk along; people might have thought his legs were simply weakened, not paralyzed.

With a public that was gripped by fear of this terribly frightening disease, the news of a preventive vaccine was greeted with great excitement, only to have that excitement dashed by a manufacturing error that resulted in many children being sickened by the vaccine and even causing the deaths of a few. While the situation was remedied and other vaccines were quickly ushered forward, the experience with polio explains the medical world into which Archie Cochrane, known as the father of evidence-based medicine, entered. Cochrane advocated for a new way of managing patient care, one that relied on clinical evidence as viewed through the lens of the physician, who was also expected to factor in the specific needs of the patient when recommendations for treatment were made.

The chapter begins with a description of the devastating effect polio had on American children and how two different physicians—Jonas Salk and Albert Sabin—sought to find a way to prevent the disease. Though the two men were never able to reconcile their different approaches, a polio vaccine did emerge and has saved the world from more outbreaks of this terrible scourge. As doctors fought to create better vaccines and improved treatments, they began to realize the importance of a systematic approach to medicine. The chapter concludes with an explanation of Archie Cochrane's campaign for evidence-based medical practices.

Jonas Salk (1914–95) is best known for creating the killed-virus polio vaccine. *(Centers for Disease Control)*

POLIO OUTBREAKS CREATE GREAT FEAR

Small localized paralytic polio epidemics began to appear in Europe and the United States around 1900 and spread to Australia and New Zealand during the next decade. These outbreaks of polio in the early 20th century were alarming, but not because the disease was new. Polio had actually been around for a long time. In the past, however, its pattern had been very different. The disease spreads through fecal matter, so children who lived

---

POLIO PRESENTS IN VARIOUS FORMS

Like many other illnesses, not every form of polio is the same. The disease can present itself in any of the following forms:

- **Abortive polio** is the mildest form of polio, as referred to by the Centers for Disease Control (CDC). Up to 95 percent of people who have this polio never suspect that they actually have it. For them the disease is limited to a flulike illness, often with upper respiratory involvement (sore throat, etc.).
- **Nonparalytic polio,** the next level of the illness, generally involves *aseptic* meningitis. According to the CDC, 1 to 5 percent of people with this type of illness show neurological symptoms such as light sensitivity and neck stiffness.
- **Paralytic polio** occurs in 0.1 to 2 percent of cases and results in permanent paralysis of the limbs, usually the legs. Of this group, 5 to 10 percent died from complications. In some cases, paralysis eventually moved to the respiratory muscles, so iron lungs were used to keep them alive. (An iron lung is a contraption used to help people breathe. Some people had to remain in these for months at a time.)

in the 19th century and earlier when sanitation standards were lower became immune to the bacteria through early exposure. By the 20th century, improvements in sanitation meant fewer children were exposed to the illness, so when they did encounter it, they tended to be older and the results were more severe. (See the sidebar "Polio Presents in Various Forms" on page 82.)

By 1950, physicians started noticing a change in the virulence of the disease. Polio was primarily affecting children; many died

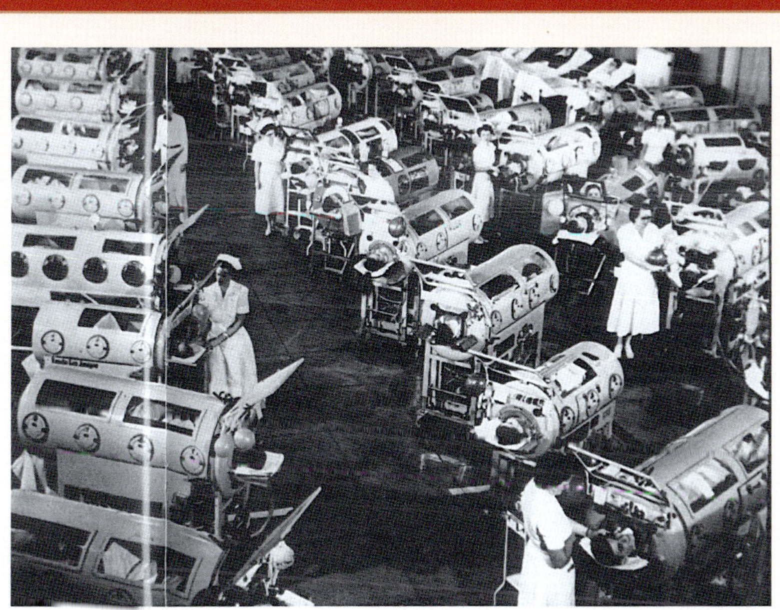

Iron lung ward of Ranchos Los Amigos Hospital, Downoy, California, ca. 1953 *(USDA)*

The polio virus enters the body via the intestinal tract and then moves to the bloodstream. If it goes on to attack the nerves, the patient may suffer one of the two more serious forms of polio.

of it or, if they survived, they might be paralyzed for life. The outbreaks occurred primarily in the summer, and there was a great fear that public swimming pools were the environment where the disease was most likely to spread. Parents fled the cities with their children once school was out and kept them out of public pools to try to eliminate exposure. In 1952, the United States encountered the worst polio epidemic in the nation's history. Of nearly 58,000 cases reported that year, 3,145 died and 21,269 were left with mild to disabling paralysis.

SALK AND SABIN

Because the results were so devastating, the public was desperate for answers, and scientists were eager to comply. Facing such a major challenge, several groups took on the task of looking for a cure. Ultimately, two well-respected scientists developed methods that worked. The biologist and physician Jonas Salk eventually set the pace by creating a dead virus that worked in a vaccine. However, when problems with that immunization process occurred, Albert Sabin stepped in with something safer and equally effective.

Jonas Edward Salk (1914–95) was a New Yorker who began his medical career in the department of epidemiology at the University of Michigan, where he had accepted an invitation to join a colleague Thomas Francis. Their first joint project was to create a vaccine against influenza. In 1939, Francis and Salk were given a grant to explore this possibility. During World War I, an estimated 20 million people died from the flu in a single year (1918). As the international scene pointed toward the likelihood of World War II, the government wanted something that might protect people from a repeat of the horrors of the influenza virus. By 1943, Salk and Francis had created a successful vaccine that worked for two of the most common types of flu, and their ability to work well together came in handy later on.

In the mid-1940s, Salk moved on to the University of Pittsburgh. He had begun working on a polio vaccine, and papers he had written on the subject came to the attention of the head of the

National Foundation for Infantile Paralysis. As a result, the organization committed major funding to Salk's efforts. Salk based his work on the efforts of those who had preceded him. Other scientists had learned that the virus is taken in via digestion and then resides in the bloodstream for a time before invading the nerve tissue. With this understanding, Salk began experimenting with ways to attack the virus before it moved beyond the blood. Most vaccines used weakened viruses to stimulate antibodies to fight disease, but Salk felt that polio was too dangerous to do this and wanted to use a dead virus. However, he needed to find a way to evaluate when the virus was dead enough not to cause infection, but not so inert that it was totally inactive. To determine this, he worked with monkeys to assess various levels of the killed virus, evaluating the quality of the vaccine based on the survival of monkeys who had been exposed to the illness. By 1952, Salk felt he had an answer and put out an announcement about what he had discovered. This ran counter to the accepted professional practices of scientists of the time who were expected to present the facts to other scientists in a scientific journal before announcing a cure, and Salk was greatly criticized for not following protocol. Salk was unperturbed; he felt the country was in the midst of a crisis and needed the information, so in 1954 he began preparing to inoculate more than 400,000 children in a clinical trial. The old team of Salk and Francis reunited for this event, with Thomas Francis serving as director of the trials. More than 1 million children ages six to nine were vaccinated. One-third received the prepared vaccination; one-third received an injection that was actually a placebo; the final group was the control group and received nothing at all.

By 1955, America knew that Salk's vaccine was a success. It was formally announced that for 90 percent of the clinical participants Jonas Salk's vaccine was very effective and safe—these statistics went a long way toward easing the public's worry about future polio epidemics.

Once news of Salk's success became known, there were all types of grateful responses from the worried public. Churches held

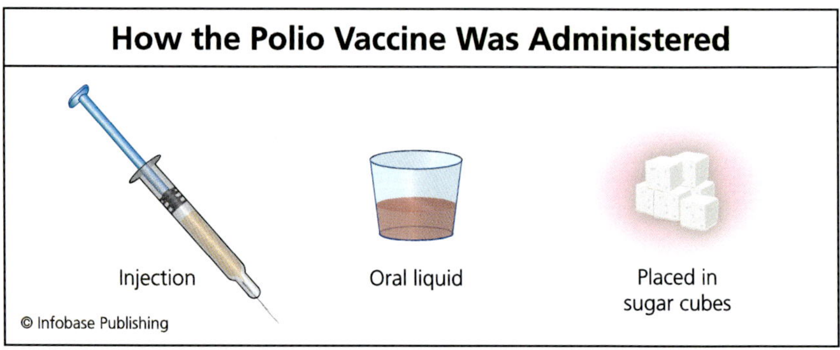

When Jonas Salk's killed-virus polio vaccine was first made available, it was given by injection. Later on, Albert Sabin developed a vaccine using a weakened form of the virus, and it could be given in liquid form or placed in sugar cubes, making it easier to administer.

special services to give thanks and communities arranged for the town bells to ring or observed moments of silence in Salk's honor. The public celebrations quieted quickly when a tragedy occurred within the first month of the vaccine's use.

Prior to the clinical trial, Jonas Salk visited each facility that was making the vaccine to train them in working with the virus, but once the trial was over and had been so successful, Salk assumed the manufacturers understood the process to be used in creating the vaccine. Unfortunately, one of the facilities made a mistake. Two hundred of the children injected came down with paralytic poliomyelitis, and 11 of them died. The public was horrified, and the government temporarily abandoned the vaccination program. The fear of polio returned.

However, the medical community was better prepared than they had been before Salk's discovery. Other scientists had been looking for cures for the disease or better ways to vaccinate, so the disruption of the vaccination program opened opportunities for other researchers. One such was Albert Sabin, a Russian who had become a naturalized U.S. citizen. Dr. Sabin believed that working with a live virus was more effective and actually less dangerous than handling a dead one, and he was at work on a vaccine that used a weakened form of the virus. In the late 1950s, Sabin conducted tests, first on monkeys and then on humans (including his own family), and by 1960 the U.S. Public Health Service approved it for manufacture.

Salk and Sabin never made peace with their two views of the polio vaccine. While both vaccines were effective, Sabin's grew to be more popular because it could be given orally instead of by injection, making it easier to use. Though Salk is more commonly associated with developing the vaccine—and he certainly did so

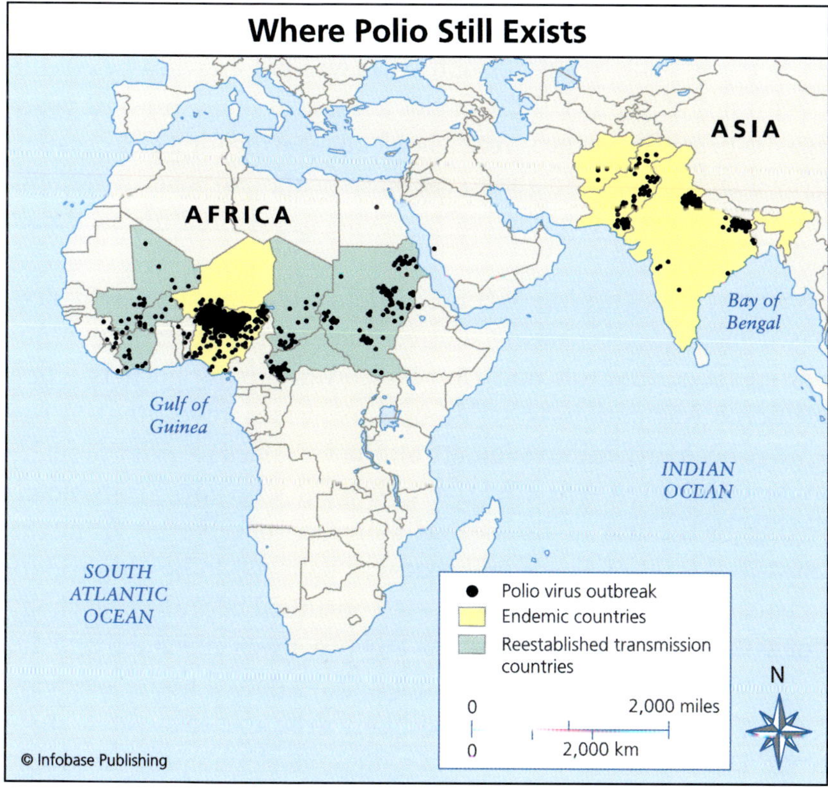

A global effort has been made to eradicate polio, and cases have decreased by over 99 percent since 1988 (from an estimated more than 350,000 cases to only 1,997 reported cases in 2006). Today, only four countries in the world remain polio-endemic (India, Nigeria, Pakistan, and Afghanistan).

first—Sabin's vaccine has proven to be more useful to the world. After these early efforts, countries around the world have tried to vaccinate an increasing proportion of their populations. In 1985, the World Health Organization (WHO) began a worldwide effort to eradicate polio by 2000, and no new case of polio has been reported in America since 1991.

A NEW METHOD FOR MEDICAL DECISION MAKING

For those who read the fine print in the financial stimulus package signed into law by U.S. president Barack Obama in 2009, one of the health care provisions for which money is being set aside traces back to a philosophy that was put forward by Archie Cochrane, who pushed for "evidence-based medicine" 40 years ago. In 1972, Cochrane wrote a book advocating for a clear system to evaluate a condition and to use statistical reports on treatments to determine what method of cure was best suited for each particular patient.

Since the beginning of time, healers have practiced evidence-based medicine, repeating the methods that worked and eliminating those that did not. However, even during periods when medicine was patient-centric (as Hippocrates advocated), it has always been difficult to evaluate all the variables in order to really understand what was effective and what was not. From bloodletting to the use of the bezoar stone (a rocklike bit from the intestinal tract of an animal that was believed to be an antidote to poisons), certain practices were continued because everybody knew they were effective, though any proof was anecdotal and did not factor in whether a person might have gotten better anyway.

Surgeons were particularly unlikely to be able to evaluate their work. During the era of barber-surgeons, most who performed surgery came into town, treated patients, and moved on, so they never knew the outcomes of their procedures. In addition, there were many misunderstood elements in the surgical process. Surgery might result in infection because of unsanitary tools, an unclean environment, or a botched job, and few took the time to adequately

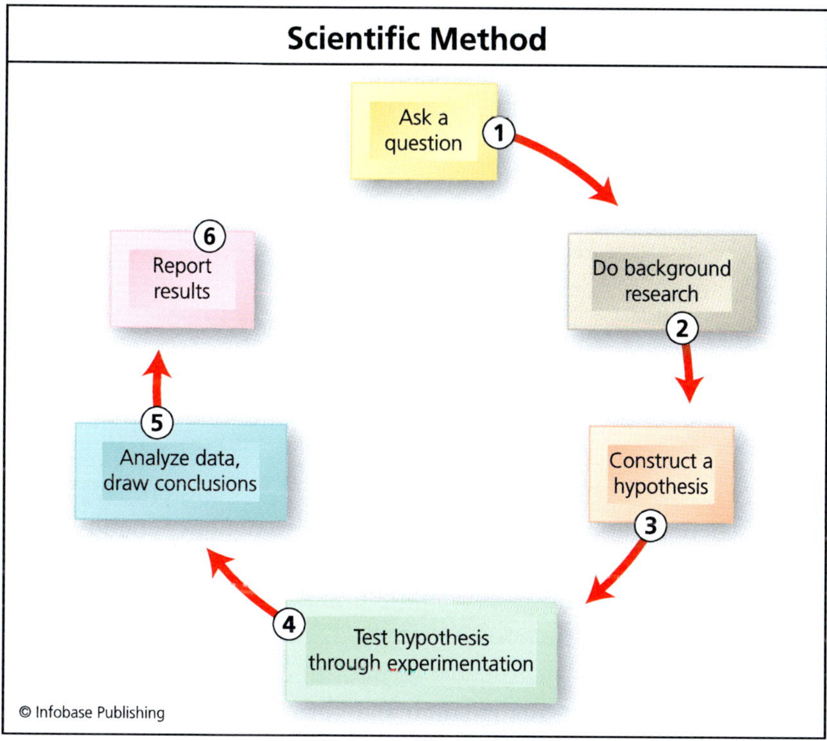

Archie Cochrane believed strongly that the only way to judge medical effectiveness was to apply the scientific method to the process of evaluating how to treat a disease. By looking for specific evidence of whether or not a particular treatment worked, Cochrane believed that physicians could make better decisions. He called this evidence-based medicine.

assess what had happened. As a result, the lessons learned were unclear.

In Michael Kennedy's book *A Brief History of Disease, Science, and Medicine,* he cites a 1973 study of surgical procedures in Vermont that clearly explains what happens. The researchers found a vast variation in the rate of common surgical procedures. In some communities, nearly 60 percent of children had tonsillectomies; in other parts of the state, the number of tonsillectomies hovered at 7 percent. There seemed to be no particular guideline for deciding which children got tonsillectomies, and amazingly, there seemed to be no noticeable difference in the medium-term health of the children. Those who had the procedure fared about as well as those who did not.

The focus of evidence-based medicine is to apply the scientific method to medical practice. A clear definition of it might read: Evidence-based medicine is the conscientious, explicit, and judicious use of current best evidence in making decisions about the care of individual patients.

There is always some concern that by the book medicine discounts the feelings of patients or the best judgment of physicians. Others criticize it because they worry that evidence-based medicine will be driven by insurance companies who only will pay for the cheapest treatments. At present, local practitioners may order tests and treatments that are not effective. Most administrators point out that if properly practiced, evidence-based medicine will identify and apply the most effective interventions to maximize the quality and quantity of life for individual patients; this may raise rather than lower the cost of patient care.

Medicine will always involve the art of evaluation and decision-making. The emphasis on evidence-based medicine involves looking at the statistics. This information may reveal how best health care dollars should be spent so that people don't waste their time—and possibly sacrifice their health—by undergoing treatments that do not work well.

ARCHIE COCHRANE (1908–1988): ADVOCATE FOR EVIDENCE-BASED MEDICINE

Archie Cochrane is best known for *Effectiveness and Efficiency: Random Reflections on Health Services,* his influential book that was published in 1972. The principles he set out were straightforward: He suggested that because resources would always be limited they should be used to provide equitably those forms of health care that had been shown in properly designed evaluations to be effective. In particular, he stressed the importance of using evidence from clinical trials, because these were likely to provide

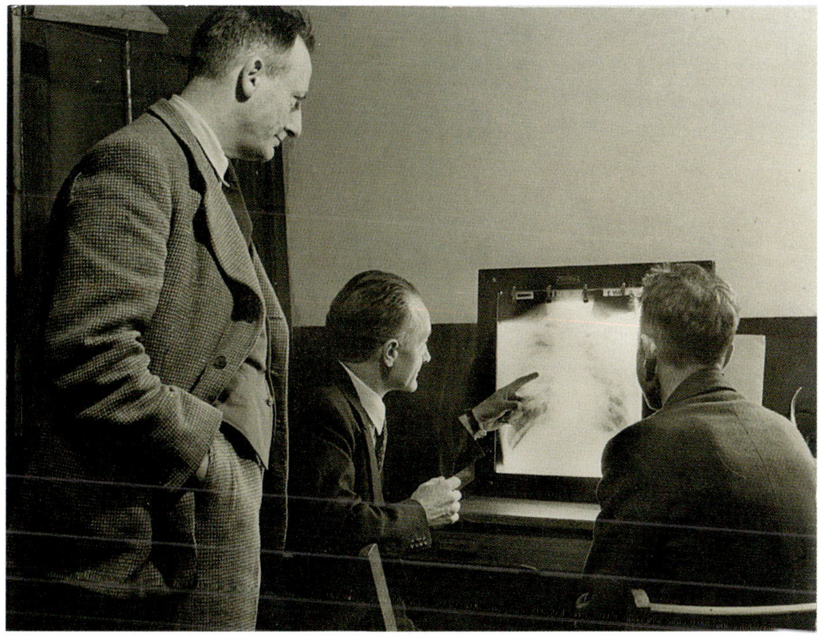

Archie Cochrane *(Cardiff University Library, Cochrane Archive, University Hospital Llandough)*

much more reliable information than other sources of evidence. Cochrane's simple propositions were soon widely recognized as very important.

Cochrane's personal experiences had a great deal to do with his recognition of the importance of new methodology. During World War II, he was held as a prisoner of war for four years, but because he was a physician he also served as the community doctor, which gave him an opportunity to observe and consider patient care. He noted that in taking care of patients with tuberculosis, he had no real knowledge of what worked and what did not. "I knew there was no real evidence that anything we had to offer had any effect on tuberculosis, and I was afraid that I shortened the lives of some of my friends by unnecessary intervention."

In 1987, the year before Cochrane died, he wrote that he saw signs of progress. He noted that a systematic review of clinical trials of care during pregnancy and childbirth was "a real milestone

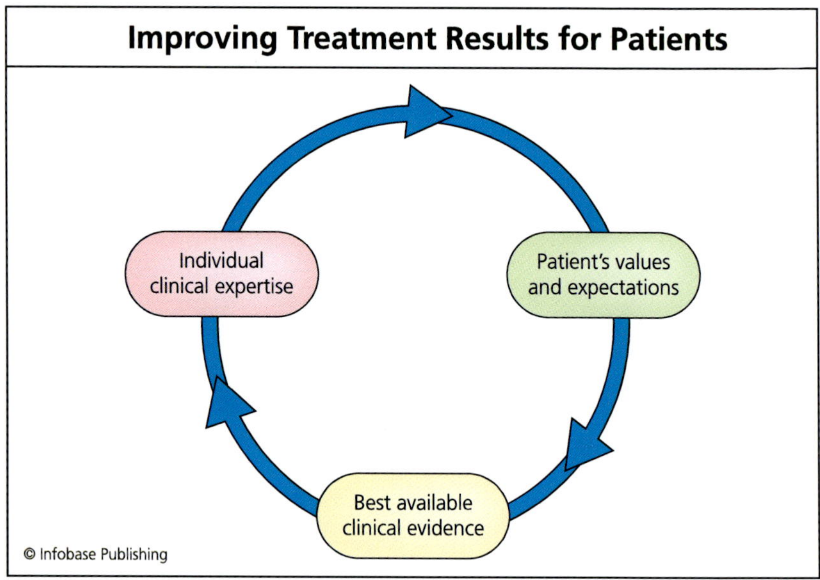

Today's medical professionals know that the three most important elements in optimizing a patient's care include a combination of the best available clinical evidence, an understanding of a patient's own values and expectations, and the physician's clinical experience and expertise guiding the process.

in the history of randomized trials and in the evaluation of care," and suggested that other specialties should copy the methods used. His encouragement, and the endorsement of his views by others, led to the opening of the first Cochrane Centre (in Oxford, Britain) in 1992 and the founding of the Cochrane Collaboration in 1993.

Cochrane would be the first physician to readily acknowledge that practicing medicine involves science and statistics—but also human kindness and consideration for the specific needs of the patient. In his autobiography he wrote of a wartime experience from which he drew an important lesson:

Another event at Elsterhorst had a marked effect on me. The Germans dumped a young Soviet prisoner in my ward late one night. The ward was full, so I put him in my room as he was moribund and screaming and I did not want to wake the ward.

HOW MEDICAL SCHOOLS PRESENT EVIDENCE-BASED MEDICINE

Evidence-based medicine is described as requiring new skills from the clinician. He or she needs to be adept at searching through relevant literature and applying formal rules of evidence. At the same time, the patient's needs and the physician's experience must be factored in. Most schools stress that any recommendations taken from evidence-based medicine must be applied by a physician to the unique situation of an individual patient. Sometimes there is no reliable research evidence to guide decision-making, and some conditions are rare enough that there is no way to do large studies.

THE STEPS IN THE EBM PROCESS

This method emphasizes the importance of the evidence, but the process begins with the patient.

1. Start with the patient—a clinical problem or question arises out of the care of the patient
2. Construct a well-built clinical question derived from the case
3. Select the appropriate resource(s) and conduct a search
4. Appraise that evidence for its validity (closeness to the truth) and applicability (usefulness in clinical practice)
5. Return to the patient—integrate that evidence with clinical expertise, patient preferences, and apply it to practice
6. Evaluate your own performance with this patient

Schools that employ this methodology in the medical profession intend to strive for the highest standards for patient care.

I examined him. He had obvious gross bilateral cavitation and a severe pleural rub. I thought the latter was the cause of the pain and the screaming. I had no morphia, just aspirin, which had no effect. I felt desperate. I knew very little Russian then and there was no one in the ward who did. I finally instinctively sat down on the bed and took him in my arms, and the screaming stopped almost at once. He died peacefully in my arms a few hours later. It was not the pleurisy that caused the screaming but loneliness. It was a wonderful education about the care of the dying. I was ashamed of my misdiagnosis and kept the story secret.

While Archie Cochrane is not well known to laypeople today, his contributions should not be underestimated. As governments and medical administrators in the 21st century wrestle with how to improve the quality of medical care, many of the principles outlined by Cochrane will be the underpinnings of these discussions.

THE IMPORTANCE OF CLINICAL TRIALS

Today's clinical trials are an important part of the medicine approval process and are carried out using strict protocols that adhere to accepted standards of safety, patient care, and data interpretation.

Early history offered no systematic method for the study of medicines or treatments. Any form of testing that was conducted was totally unscientific and anecdotal. Someone may have been recovering from smallpox and, if the physician performed bloodletting and the patient continued to improve, then bloodletting was considered curative. Today, scientists know that medicines need to be standardized, and different delivery methods (taken orally, injected, etc.) can make a difference in the outcome. A herbal medication might have been reported to be helpful, but the next healer had no way to duplicate the medicine or the delivery method so they had no real way to test or guarantee results.

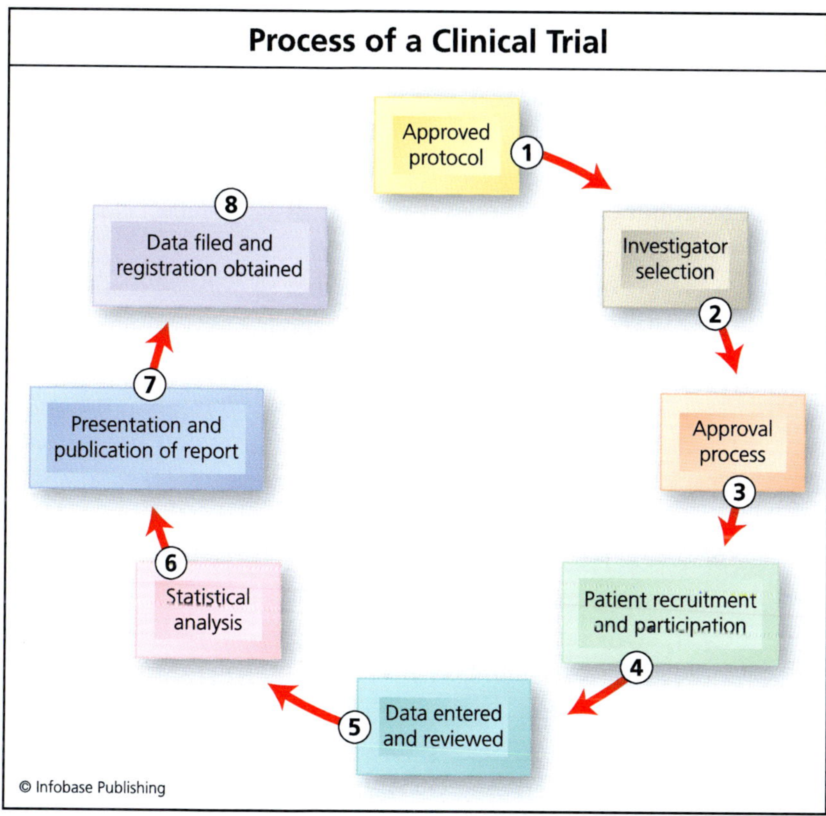

Clinical trials are a vital part of the medical system because they provide an unbiased evaluation of whether a particular treatment or medication is effective. To make certain that trials reflect accurate information, they must be carefully supervised and all must abide by the same protocol so that the results are valid.

Physicians today understand the importance of comparing two treatments, though they readily acknowledge the difficulty of reducing the variables. Early medical practitioners were hard-pressed to take care of patients; the thought of comparing treatments would have been almost unthinkable. Occasionally an opportunity occurred by accident. One of the first to be able to compare two treatment methods was Ambroise Paré, a surgeon in the 16th century. Paré was on a battlefield treating wounded soldiers. The method at that time involved pouring hot oil over the injury. When he ran out of oil, he created his own concoction of turpentine, rose oil, and egg yolk. After a fitful sleep caused by worry about his patients, Paré rose early and went to check on his patients. To his great surprise, he found that those who had received his concoction fared better than those who had been treated with hot oil. The opportunity for direct comparisons such as this was rare.

James Lind is often considered the father of clinical trials, as he was the first to introduce control groups into his experiments. Lind was very interested in finding a way to help sailors avoid scurvy (a debilitating illness that caused spongy gums, bleeding, and tooth loss). Some sea captains had begun to note that when fresh fruits and vegetables were brought on board, the men were less likely to become ill and could remain active workers. In 1747, Lind set sail on the *Salisbury*. He provided all of the men with the same general diet but some received additional items, including cider, elixir vitriol, vinegar, seawater, nutmeg, oranges, and lemons. Those who received the fruit and foods with vitamin C (including the cider and vinegar) fared better. Because citrus products were expensive and did not last well, the British navy resisted adding lemon juice to the provisions for seagoing vessels. Later, they determined that lime juice worked just as well as lemon juice. It was cheaper, so lime juice eventually was added to the Navy supply lists. (This is why British sailors, and later the British in general, were called limeys by Americans.)

By the early 19th century, scientists and physicians began to conduct clinical trials more frequently, and they started experimenting with the design of the trials. They realized the psychological effect of providing everyone with something and for the first time placebos were used (1863). (Placebos have no pharmacological effect but by providing everyone with some type of substance scientists are better able to evaluate whether simply giving the patient something is curative in certain circumstances.)

Most clinical trials now involve the use of placebos, but patients do not know what they are getting—an actual medication or a substitute. In 1923, this process was further amended to randomize who received what so that there was no way to play favorites with one group or another. Today, studies usually are also double blind. In double-blind studies, the medical professionals as well as the patients are not informed of who is getting what. This means that medical professionals conduct themselves similarly with all patients, and it helps avoid the problem of looking for particular results.

Since 1945, government regulators in many countries have struggled with how to structure clinical trials so that there is a balance between a scientific need to know and patient safety. For example, if physicians are relatively certain that a certain medication is helpful to cure or improve a disease, there are ethical implications to placing certain people in the control group receiving the placebo. Today, clinical trials are part of standard procedure and are required as part of the government review process, but there are guidelines to safeguard patients. Informed consent—explaining to patients the exact nature of the study and the risks involved—is always required for clinical tests run in the United States.

CONCLUSION

Polio was a frightening illness that terrified the public. Jonas Salk and Albert Sabin worked separately to create vaccines that could be used effectively, and today in most of the world the threat of polio is just a dim memory.

On the heels of the progress with polio, physicians began to realize the importance of systematizing care. Archie Cochrane and his work toward an evidence-based practice of medicine increased the level of professionalism for practitioners, and today clinical trials and a better examination of treatment efficacy are important aspects of the medical profession.

In 2009, there is renewed interest in evidence-based medicine and further pinning down what is effective and what is not. President Barack Obama has indicated that part of the solution to halting spiraling health care costs involves reexamining what constitutes effective medicine. He has stated that he wants the government to undertake comparisons of therapeutic data based on a cross-section of the population. By employing better technology to study health care, President Obama hopes the country will be better able to evaluate what treatments are effective for which patients, taking into account their ages, genders, genetic profiles, and even social habits. This is the 21st-century version of exactly what Archie Cochrane had in mind.

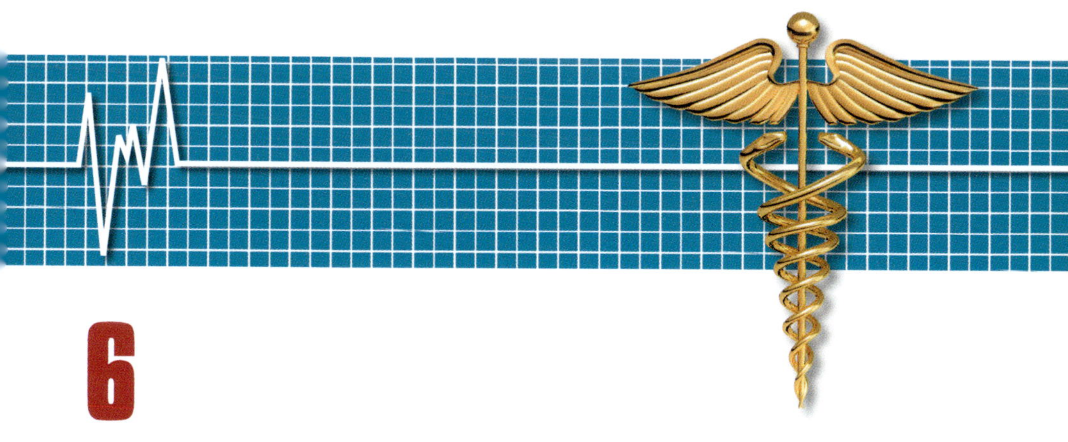

6

More Changes Brought about by War

The disastrous injuries that soldiers undergo during war always present new challenges for medical professionals. Evolving types of weaponry create new and different wounds; changes in the locales of wars mean climates differ, which often introduces new forms of bacteria or causes bacteria to behave differently. The needs of the soldiers also change. Injuries sustained in wars fought before antibiotics were more likely to result in death from infection (if soldiers survived they had shorter life spans so their needs to become contributing members of society seemed less intense). Today, the conflicts in Iraq and Afghanistan, for example, have resulted in relatively low fatalities, but more men and women are returning home with injuries and wanting and expecting to join the mainstream world again.

The period of medicine described in this book begins in 1840, so before going forward to discuss the advances of the 20th century it is important to step back to consider the circumstances of soldiers during the Civil War. Medical care at this time was extremely primitive. Physicians had no access to anesthesia near the battlefield nor did they know of Joseph Lister's work in understanding the importance of a sterile environment for surgery. Physicians would take over a home to create a temporary hospital, and they did the best they could with the horrors that confronted them. The problem of a wounded limb frequently had to be solved by amputation, and luck dictated whether a soldier's wound became infected—a medical condition for which physicians had no effective remedy.

By World War I, scientists knew that bacteria caused infections, but they did not yet have a cure available. The introduction of sulfa drugs and penicillin (see chapter 5) had a great impact on soldiers' survival, and between 1919 and 1939 the expanding use of technology and improved scientific techniques created a higher level of medical care in general. There were better forms

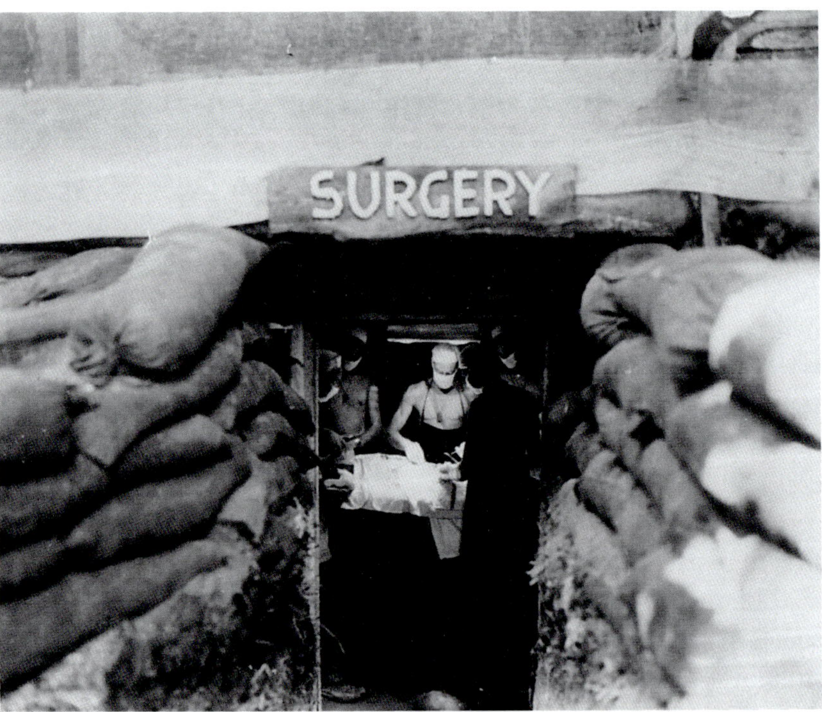

In an underground surgery room, behind the front lines on Bougainville Island in the Pacific, an American army doctor operates on a U.S. soldier wounded by a Japanese sniper, December 13, 1943. *(National Archives and Records Administration)*

of treatment, blood could be transported to most battlefields, and the medical professionals on the ground were constantly working to improve the methods of patient triage. Since that time, soldiers have benefited from the computerization of diagnostic processes and much more rapid processing of data.

Perhaps there is no more significant change for soldiers in the last 150 years than in the different options given soldiers today who lose a limb. During the Civil War, a solider was lucky to be given a stump of some sort to replace a missing body part. Today, medical and bioengineering professionals are opening new possibilities in the world of prosthetics.

Social acceptance has been part of this change. In the aftermath of World War I, the injured who

Civil War prosthetic *(Ty's Journey)*

returned missing a limb or with a decided limp were referred to as cripples, but a heightened awareness brought about by social activists who fought on behalf of these seriously wounded soldiers has now created an environment where it is not even noticeable that someone has a prosthetic leg. Also, federal law dictates that buildings must be accessible to the handicapped, and it is not unusual to see people with prosthetics entering races or other competitions of physical prowess.

War has also brought about improved methods in other areas of treatment. Blood management, emergency room procedures, the use of vaccines, and better pain management all benefited from ideas that were born on battlefields.

ADVANCES IN PROSTHETICS

The American Civil War brought about the need for a field of prosthetic medicine. It is reported that there were at least 30,000 amputations on the Union side alone, but at the time the best doctors could do was provide a very primitive limb substitute.

An advance in another area, the development of anesthesia, created an opportunity for better surgical procedures. Later in the 19th century, with the introduction of anesthesia, doctors could perform surgeries that were more detailed and took longer, because the patient could be adequately anesthetized. As a result, physicians began to develop ways to operate on the patient's stump in such a way that a prosthetic device could more easily and more comfortably be fitted on to the person's limb. As surgeons improved their techniques and learned successful ways to prevent infections, their success rate improved. In addition to better prosthetics, another important development that occurred in the early 20th century was a social movement inspired by two efficiency experts who encouraged acceptance of people with disabilities in the workplace. (See the sidebar "Pioneers in Helping the Handicapped" on page 107.)

World War II was the next occasion that encouraged the improvement of prosthetics. In 1945, the National Academy of Sciences, an American government agency, established the artificial limb program in response to the influx of World War II amputees and for the purpose of advancing scientific progress in artifi-

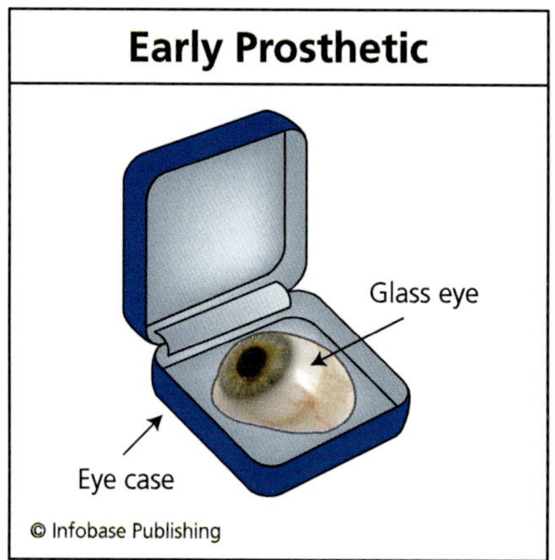

People have always wanted to look as normal as possible. A person who lost an eye preferred to have a prosthetic. Glass eyes were created so that a thin thread could hold the piece in; today's acrylic materials are much lighter and easier to wear.

cial limb development. Since then, advances in areas such as materials, computer design methods, and surgical techniques have helped prosthetic limbs to become increasingly lifelike and functional.

As technology has progressed, lighter, stronger materials have been developed for use in prosthetics, and scientists have increased the comfort and usefulness of the devices. Today, electronics can be added so that an amputee can make the limb function in a relatively normal way. Innovation has also led to making the prosthetics look more natural. Today's amputee may actually have more than one prosthetic limb, with some better-suited for certain activities.

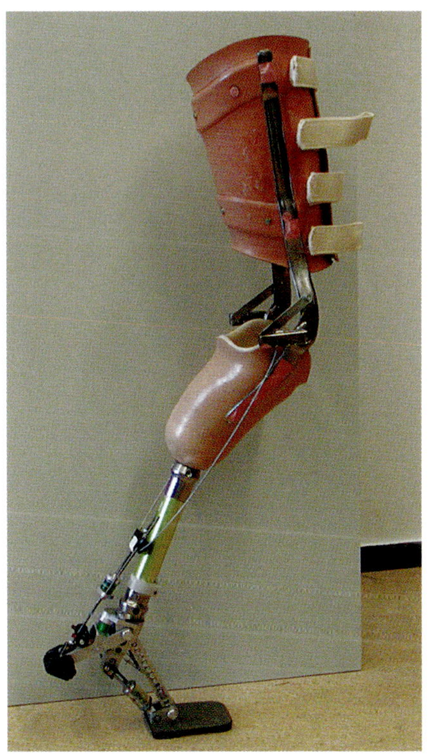

Prosthetic leg *(UAF Engineering)*

Unlike the Civil War era when any stump would do for any patient, amputees today are carefully measured and their devices are custom built and fitted for maximum comfort and usability. Advanced lower extremity prostheses are equipped with a variety of mechanisms that help them to move naturally as a patient walks or runs. A prosthetic knee is particularly difficult to engineer, as it must constantly adjust to allow for normal walking, standing, and sitting. If the person requires a full artificial leg, then modern ones are created with a computer-controlled knee that automatically adapts to adjust to the patient's walking style.

The current advances in prosthetics involve using a patient's muscles to command the prosthetic limb. Because muscles generate

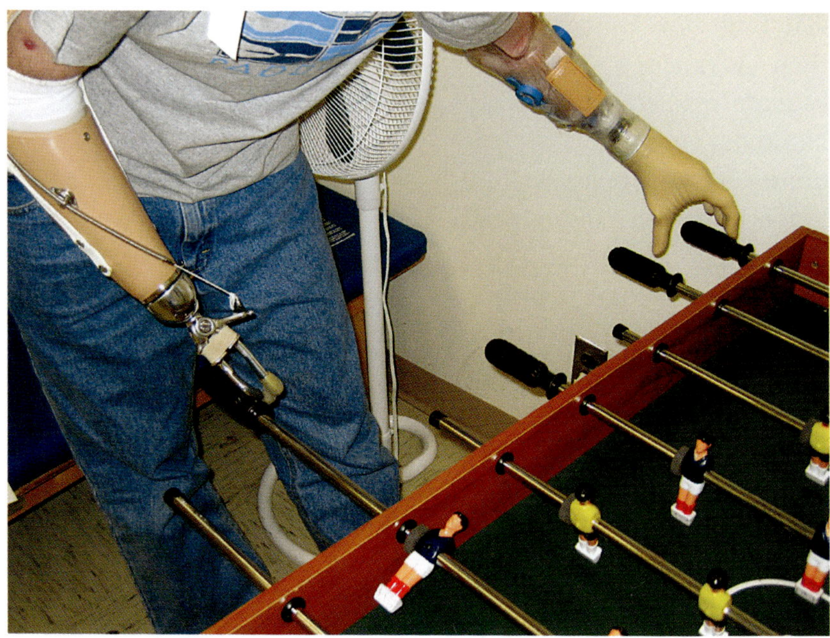

Playing football with prosthetic arms *(U.S. Army)*

small electrical signals when they contract, electrodes placed on the surface of the skin can measure muscle movements. Although no buttons are physically pressed by the muscles, their contractions are detected by the electrodes and then used to control the prosthetic limb. These prosthetics are called myoelectric.

Another type of surgery that benefited from the lessons learned by surgical amputation and the creation of prosthetics had to do with another area where there is a great wish for advancement—that of limb replantation. The first successful case of a limb being reattached occurred in 1962 when an accident severed the arm of a 12-year-old boy. The child and his arm, which had been cut off just below the shoulder, were rushed to Massachusetts General Hospital where surgeons grafted the arm to the boy's body. Some months later the nerves were reconnected, and two years later the boy had regained almost full use of his arm and hand. Since that time, replant operations are attempted when the right conditions are present.

Physical therapy is critical after an amputation surgery. Whether a person is trying to learn to use a new arm or whether they are retraining themselves to walk with a prosthetic limb, it is a difficult undertaking, requiring several months of rehabilitation and training. Additional adjustments to the device may also be necessary. For a leg prostheses, the person fitting the prosthetic carefully monitors the walking gait of the patient and makes adjustments as necessary. In addition, therapy to keep the patient's body strong enough to manipulate the prosthesis is an important part of the process.

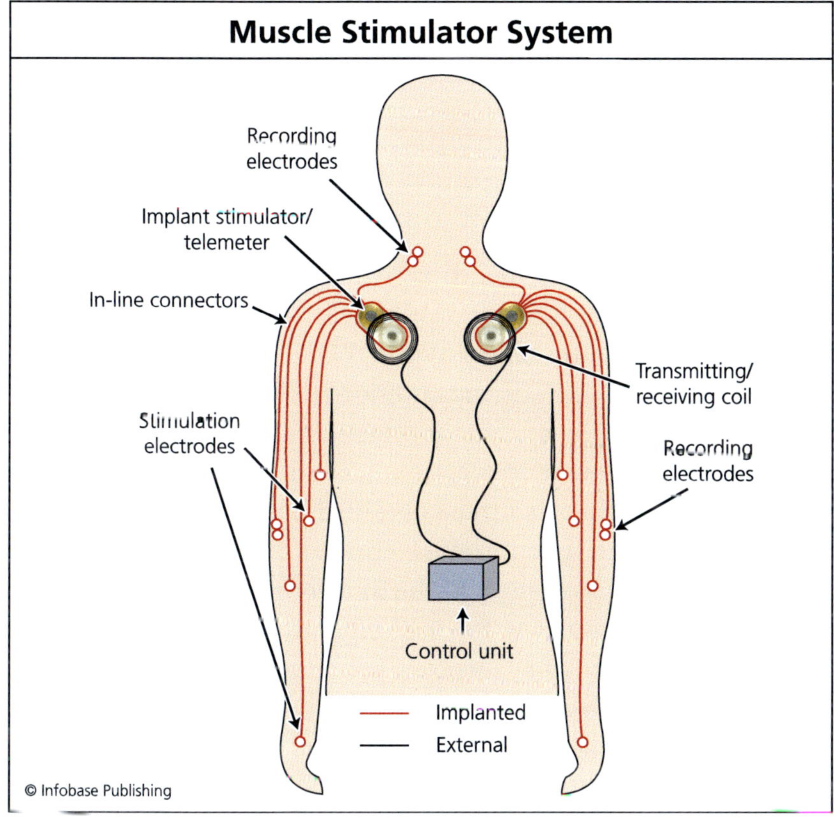

Neuromuscular electrical stimulation has been used to revitalize purposeful movement to muscles crippled by spinal cord injuries.

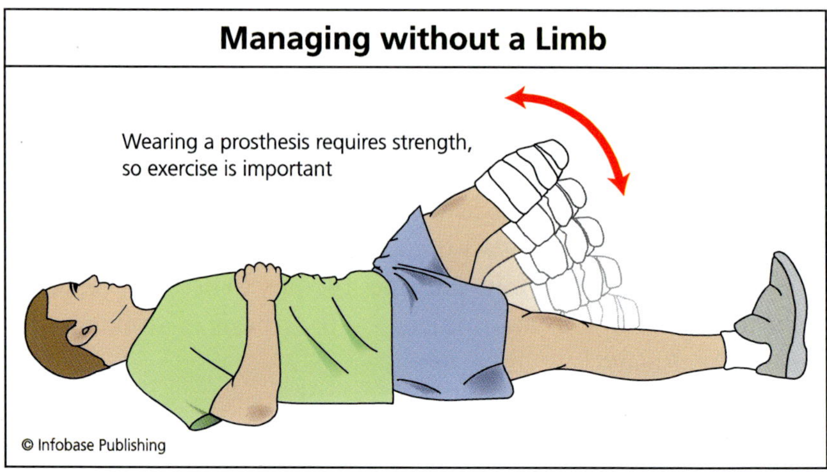

Learning to live with a prosthetic requires building up one's muscles and learning a new way of controlling one's body. Exercise and physical therapy are important steps in the process.

THE USE OF VACCINES

Wartime has proven to be a time when vaccines are tested and their use extended, often to the benefit of the general population. Before the outbreak of World War I, the U.S. Army had begun requiring certain vaccinations. Typhoid inoculations were given regularly, and the military wished for a way to vaccinate against tetanus. Protection against tetanus relied on an antitoxin serum that had to be obtained by immunizing horses. There had been no time for much clinical testing on humans, so this initially slowed the use of the antitoxin tetanus serum, and it was primarily reserved for use only on the injured. Progress by Emil Behring for the Germans eventually led to the use of a serum for all soldiers. By World War II, the antitoxin to tetanus had been transformed into a vaccine that could be used before the men entered battle. Today, children are always vaccinated against tetanus, and people are vaccinated against typhoid on an as-needed basis.

PIONEERS IN HELPING THE HANDICAPPED

Frank and Lillian Gilbreth are best known for their motion study methods that involved helping American industry. However, they became very concerned about the disabled veterans returning from World War I. The Gilbreths demonstrated how handicapped workers could become productive members of society. Their book *Motion Study for the Handicapped* (1917) was the first to deal in depth with occupational rehabilitation. (Lillian Gilbreth is perhaps best known as the mother of 12 from *Cheaper by the Dozen,* written by two of her children, Frank Gilbreth, Jr., and Ernestine Gilbreth Carey, published in 1948 and released as a film in 1952, which documented Gilbreth family life.)

While the majority of their work was done for major corporations looking to maximize the efforts of their workforces, Frank and Lillian advocated that disabled workers should be included in corporate workplaces. They felt strongly about the importance of matching the job to the workers, and they conducted studies to identify the types of tasks that handicapped workers could best perform. They also realized the importance of rehabilitating injured soldiers, and they advocated the disabled be given special training. The results of their studies were impressive enough to attract the attention of the government, and their work was incorporated in the Vocational Rehabilitation Act of 1918 passed by Congress to meet the needs of disabled veterans.

After Frank's untimely death in 1924 from a heart attack, Lillian continued on with their work, and she incorporated the needs of the handicapped in much of what she did. When she was asked to write a book about homemaking (as an efficient mother of 12), she added a chapter on the needs

(continues)

(continued)

of disabled homemakers. She also worked with General Electric to redesign home appliances with the handicapped in mind. In addition, Gilbreth was active with the Girl Scouts on the national level, and she encouraged leaders to take into account the needs of the handicapped—a very new thought for the time.

Frank and Lillian were unique in recognizing that the problems of the disabled were more than physical. They saw that by providing ways for the handicapped to help themselves, they helped build self-esteem—something vital to every human being.

Frank and Lillian Gilbreth with 11 of their 12 children, ca. 1920s *(Purdue Libraries' Archives and Special Collections)*

Troops today are vaccinated against many diseases, partially dictated by the part of the world where they are to serve. Some of the vaccines themselves are controversial, such as the vaccine against anthrax that sometimes has serious side effects. Nonetheless, by necessity, the military frequently has served as the testing ground for vaccines that are eventually introduced to the general population.

THE BETTER MANAGEMENT OF BLOOD

War creates an unprecedented demand for blood. The Spanish civil war was the first time that blood could be transported to the front lines. By the time the war had spread through Europe, becoming World War II, blood management was improving, and the Allied forces were backed by a well-organized blood supply. Two Americans—Edwin Cohn and Dr. Charles Drew—had revolutionized the storage and distribution of blood by creating ways that the blood plasma could be dried for transport and then prepared for transfusion with distilled water.

No matter where the war, it is difficult and costly to transport great quantities of blood to the front lines, and efforts to develop a blood substitute were intensified by the military in 1985. The surgeon Gerald Klebanoff, who served in the Vietnam War, introduced a device for autotransfusion (receiving a transfusion of one's own blood) in the military hospitals, and the Israeli army has used a product that was primarily created for hemophiliacs that stops life-threatening hemorrhaging, a major help in the treatment of injured victims.

In the 21st century, the emphasis has been on looking for ways to conserve blood. A war in a remote country like Afghanistan creates a very high cost of transfusing blood; up to $9,000 must be estimated for one unit of red blood cells because of the logistical difficulties of blood storage and transport. Today, physicians and the U.S. military are working together to find better ways of blood management. What they learn will affect soldiers as well as people at home.

MASH UNITS AND EVEN MORE ADVANCES IN TRIAGE

A physician who will long be remembered as a cardiac pioneer can also claim special credit when it comes to changes in the management of military medicine. Michael DeBakey (whose contributions to heart research and surgery are described in chapter 7) was assigned to the U.S. Office of the Surgeon General during World War II. He saw the need for a method to improve treatment of soldiers during wartime and developed a plan for what are now known as mobile army surgical hospitals. These MASH units were established as field hospitals for a mobile staff (at least one surgeon, an assistant surgeon, an anesthesiologist, an operating room nurse, and a technician) to go into the field to treat the wounded, operating similarly—but with more advanced medical techniques—to the flying ambulance corps created in the late 18th century by Dominique-Jean Larrey (1766–1842), the French military surgeon. In the 20th century, soldiers benefited from speedier treatment as well as improved medical knowledge, and as a result the statistics on survival improved. (Many people associate these units with *M*A*S*H,* the long-running 1970s television show about a hospital field unit in the Korean War.)

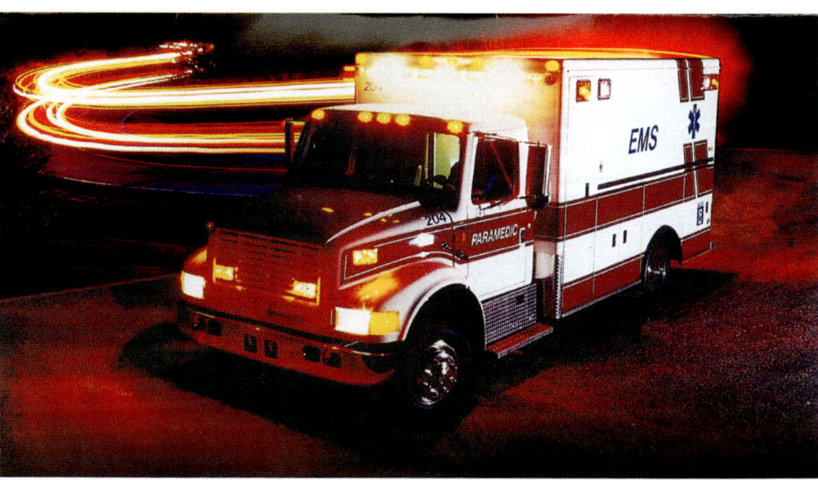

Changes in emergency medicine are one of the positive outcomes of what has been learned from war. *(St. Louis Retirement Living)*

MEDEVAC Triage Today

On the battlefield, triage methods today continue to change and improve. The most recent Iraqi War gave rise to a system where a first-responder team does what they can and makes an initial evaluation before passing the patient on to a second-stage team that is also on the battlefield. This permits medical professionals to see the greatest number of patients as quickly as possible.

Triage as outlined by MEDEVAC (Medical Evacuation) is constantly being reexamined to meet current needs. The basic triage categories (with corresponding color codes), in precedence, are as follows:

- **Immediate:** The person in this category must be seen right away or he or she may not survive.
- **Delayed:** Those categorized as delayed may have life-threatening injuries but as long as they are dealt with in six hours, they are expected to survive.
- **Minimal:** The term *walking wounded* is used for these casualties. They require medical attention but only after those in the immediate or delayed category have been stabilized or evacuated.
- **Expectant:** These patients are in very poor condition. Care should not be abandoned, but those with greater survival potential should be treated first.

The need for improved civilian emergency care is further highlighted in the sidebar "John Wiegenstein (1930–2004): Father of Emergency Medicine" on page 112.

IMPROVEMENTS IN PAIN MANAGEMENT

Another vital area that has profited from battlefield experience is the war on pain. More than 90 percent of wounded soldiers have made it off the battlefield from Iraq and Afghanistan, the highest survival rate in American history. As a result, American

JOHN WIEGENSTEIN (1930–2004): *Father of Emergency Medicine*

When John Wiegenstein graduated from medical school in 1960, hospital emergency rooms were very different from what they are today. They were primarily staffed by interns, with a rotating group of doctors on call if needed. This meant that a dermatologist might be the physician in charge when a patient was admitted in cardiac arrest. There was no formal method of emergency training, and because the hours in the emergency rooms were round-the-clock those who were actually hired by larger hospitals to staff emergency rooms tended to be those who could not get a job elsewhere. Alcoholism in these doctors was common.

When Wiegenstein was a young doctor and took his assigned rotation in the emergency room at St. Lawrence Hospital in Lansing, Michigan, he was appalled by how ill-prepared he was. During his first year, a child came in blue-black from lack of oxygen, and Wiegenstein had to perform a tracheotomy, a surgery he had never done before. The child survived, but this and other similar experiences energized Wiegenstein into thinking about how things might be changed. He began attending seminars on orthopedics,

physicians have been seeing more cases of chronic pain than ever before. This has increased the interest in spending on pain research. Since 2003, the U.S. military has been conducting a study of soldiers who are being given high-tech nerve-blocking devices that are not addictive. Early results indicate that those who get these devices early and can control their pain soon after an injury are having less chronic pain later.

Scientists are also studying the brain's involvement in pain. Physicians now know that the brain and spinal cord rewire them-

classes on surgery, and emergency medical training classes with firefighters. He saw that if done properly, emergency medicine could be a specialty for which physicians were formally trained.

In 1968, Wiegenstein organized a meeting with other physicians, and he presented them with the idea of creating an official organization of emergency room physicians, what was to become the American College of Emergency Physicians. The group started by Wiegenstein has worked at building national awareness of the need for qualified emergency care and training. At the start, they met with resistance from many professionals who felt they were invading their turf, but over time Wiegenstein's point of view that emergency physicians needed to be specifically trained for the assignment has been widely and popularly accepted. Today's emergency rooms are staffed with carefully trained specialists in the field of emergency medicine, and they have established specific agreed-upon guidelines for screening and treating patients who come in with any type of emergency.

selves in response to injuries and form pain pathways that can reactivate in later years. They are trying to develop ways ranging from pain patches to alternative medicine treatments that help rewire the painful mind-body connection.

These steps forward could also be a major boon for the general population. Chronic pain is one of the most pervasive medical problems in the United States, with one in five Americans suffering from it. This is costly both in terms of medical expenses and in lost productivity.

CONCLUSION

While it is hard to think of war having a silver lining, the medical profession and citizens alike would probably reluctantly agree that some good does come out of the horror of war. Scientists and physicians and emergency responders all rise to the occasion and provide the best care possible to soldiers. The lessons they learn are brought back home and employed in civilian facilities.

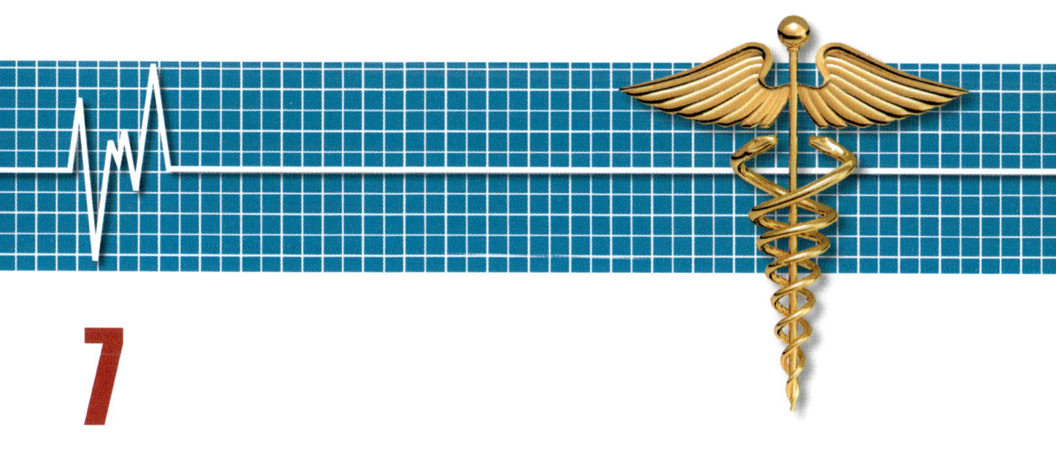

7

The Science of the Heart

Physicians today know that blood is the key to life. They can withdraw small amounts to analyze a patient's health; they can safely give a transfusion to a person who has lost too much blood; and they can separate blood to extract blood plasma or platelets or various other components in ways they could not have previously imagined. All these elements can be used in a myriad of ways for both different types of diagnoses as well as treatment.

Well into the 19th century, physicians continued to perform bloodletting, but they were beginning to note that there were times when it might be valuable to be able to add blood, not take it away. When the first human-to-human blood transfusions were performed, the failure rate was high, and progress moved in fits and starts before the Austrian physician Karl Landsteiner came up with an answer for successful transfusions.

As scientists grasped that the heart was actually a very mechanical organ, they began to think about whether it could be replaced with anything else if it became weakened or diseased. In the process, many scientists investigated the creation of artificial pumps that could mimic the heart's action and eventually began to experiment with the possibility of transplanting a good heart into a person whose heart was giving out.

This chapter highlights the progress that has been made in learning about blood and circulation and the heart in what has been a relatively short period of time. The progress made in blood transfusions and heart surgery has been remarkable, and the use of heart transplants and artificial hearts is becoming more common. Today, medical professionals stress that the best approach to heart disease is to avoid it and preach heart healthy foods, weight control, and physical exercise. This chapter also examines what scientists are currently learning about new ways to use a patient's own blood to encourage self-healing.

EARLY KNOWLEDGE OF THE BLOOD

In the early 19th century, body imbalances were still seen as the cause of most diseases, and physicians used various ways to rebalance all aspects of the body. Cathartic substances were used to empty the bowels; diuretics were given to cleanse the system by increasing urinary output, and tonics were used to stimulate a depressed nervous system. To balance the blood, bloodletting was still used for certain purposes. A patient with a fever or some type of swelling was viewed as a perfect candidate for this remedy because withdrawing blood slowed the pulse, reduced body fluids, and decreased the patient's temperature. Physicians determined that sufficient bleeding had occurred when the fever subsided, the pulse slowed, or pus developed. (Some physicians viewed pus as laudable.)

In the interests of maintaining bodily balance, physicians realized that losing too much blood could be a bad thing, and as early as the 17th century physicians began experimenting with blood transfusions. At first, progress was nonexistent. William Harvey (1578–1657), who was the first to accurately describe the circulation of the blood, also experimented with transfusing blood from animals to humans, but he learned nothing that could be passed on to others who followed him. Then in 1818, an obstetrician named James Blundell determined that transfusions needed to be within the same species, and he had some occasional successes with transfusing blood from person to person. In 1840, Blundell served as an adviser to the British physician Samuel Armstrong Lane as he performed the first whole blood transfusion to treat hemophilia. The transfusion was a success, but neither Blundell or Lane had enough knowledge about the circulatory process to tease apart the transfusion in such a way that they could analyze why it went well. Their single success was followed by many failures.

THE IMPORTANCE OF BLOOD TYPES

In order to transfer blood from one person to another, physicians and scientists began to realize that they were missing some necessary piece of knowledge that could help them make it work. They saw that even when transfusions were limited to the same species, human to human, there was no guarantee that it would work. Karl Landsteiner (1868–1943), a well-respected immunologist, was the person who found the key that permitted successful transfusions.

Landsteiner was familiar with the work of the German physiologist Leonard Landois (1837–1902) who reported that when red blood cells were taken from one species of animal and were mixed with serum taken from an animal of a different species, the red cells typically clumped and sometimes burst. Landsteiner noted Landois's observation, and by 1903 he had determined that a similar reaction occurs in some, but not all, human-to-human transfusions. When a transfused patient went into shock, became jaundiced, and often suffered *hemoglobinuria* (a condition with blood in the urine because hemoglobin levels are too high), it generally was preceded by this clumping problem.

By 1909, Landsteiner had found his answer: Human blood might all look the same when a person got a cut or suffered a wound, but it actually had variable characteristics that he referred to as blood types. He developed a way to classify human blood into

(continues on page 121)

Distribution of People with A Blood Type

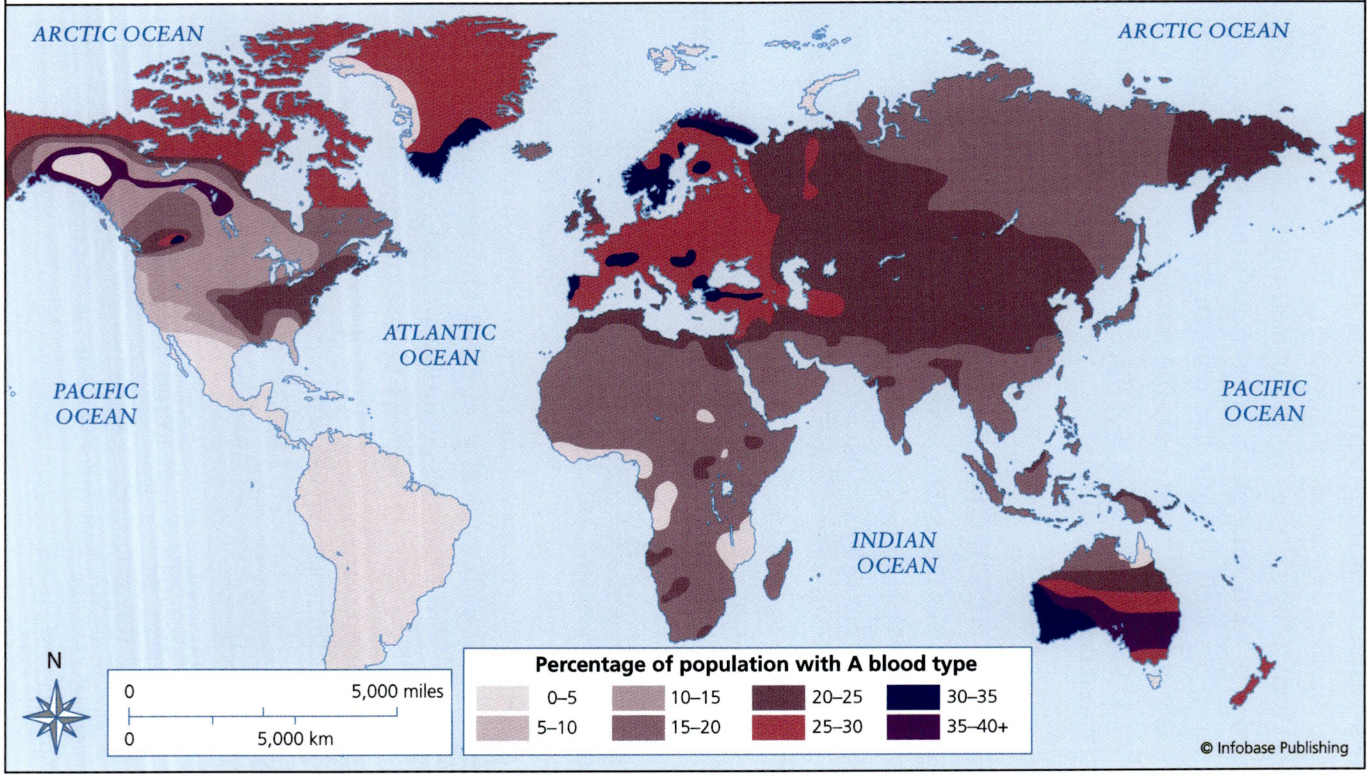

The A blood allele is somewhat more common around the world than B. The A allele apparently was absent among Central and South American Indians so there are few people with that blood type in South and Central America.

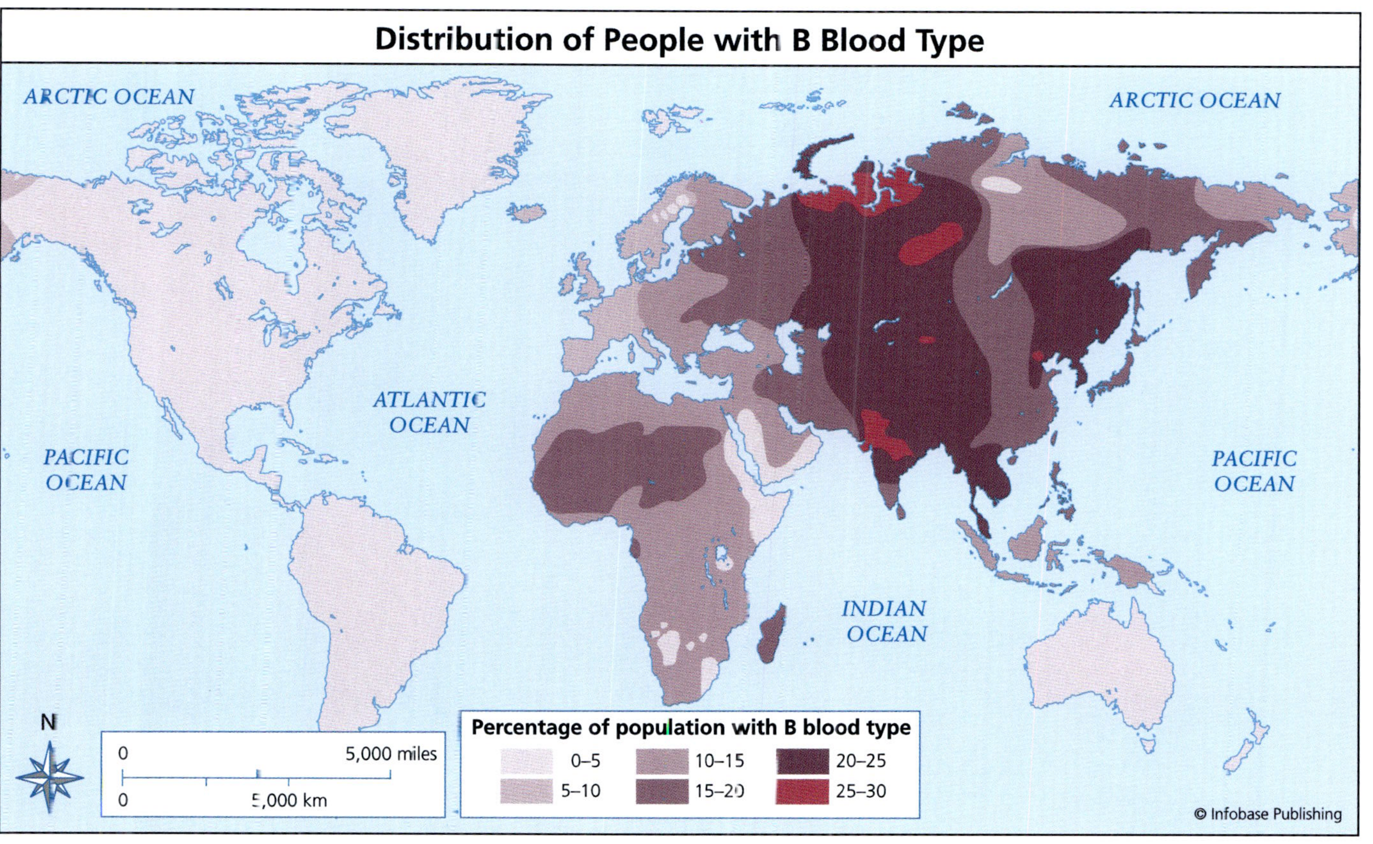

The B blood type is highest in Central Asia, and it is less prevalent in the Americas and Australia. However, there are relatively high-frequency pockets in Africa as well. Overall in the world, B is the rarest blood allele. Only 16 percent of people have it.

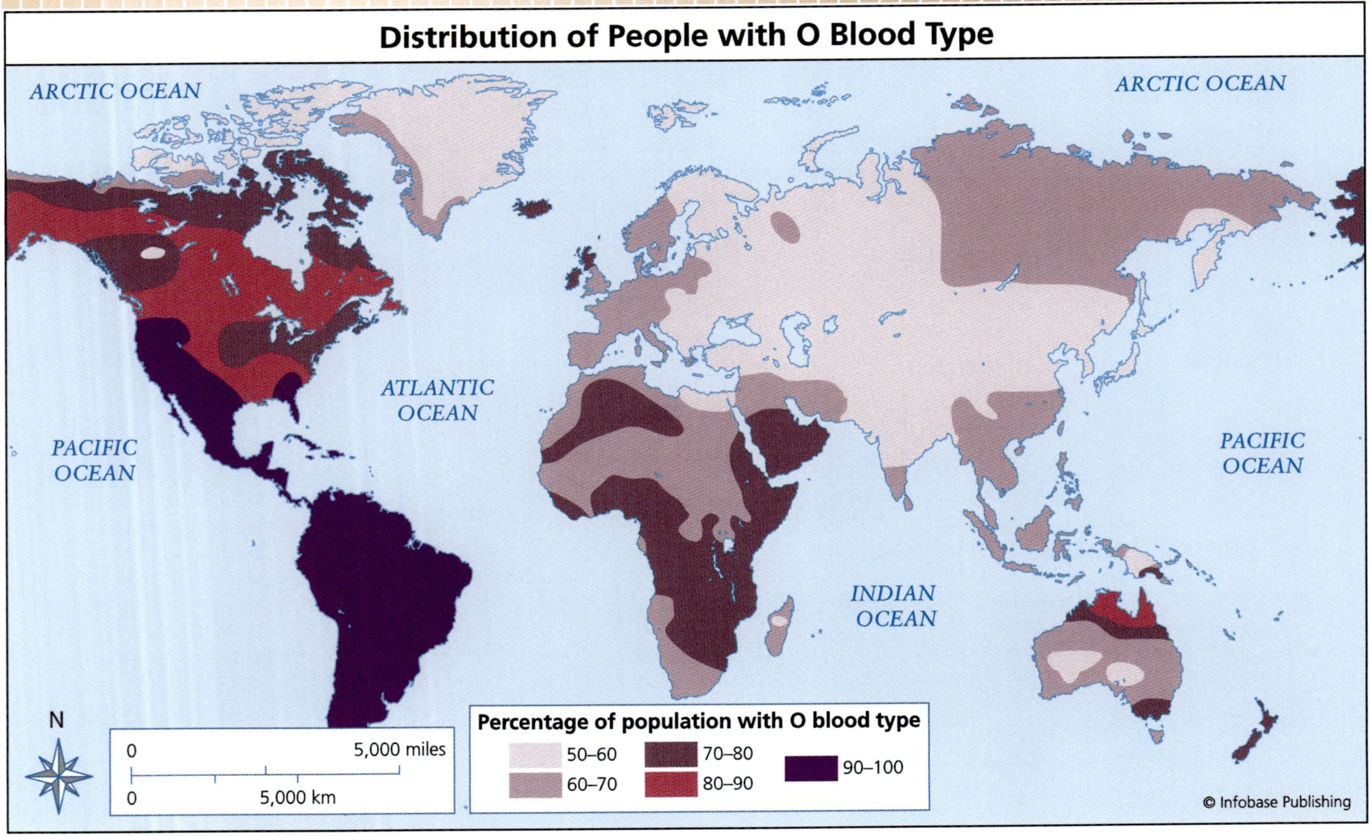

Distribution of People with O Blood Type

The O blood type (usually resulting from the absence of both A and B alleles) is very common around the world. About 63 percent of humans share it.

(continued from page 117)

the now well-known A, B, and O groups (three groups were identified initially). He went on to demonstrate that if transfusions were restricted to people of the same blood type, then all went smoothly; catastrophe occurred only when a person was transfused with the blood of a person belonging to a different group.

In 1902, two other scientists identified a fourth main blood type, AB. Shortly after this, Reuben Ottenberg performed the first blood transfusion using blood typing and cross-matching. Ottenberg also recognized that blood group types seemed to be dictated by inheritance.

Twenty years later, physicians were still encountering occasional difficulties performing transfusions. In 1939–40, Karl Landsteiner and three other scientists—Alex Wiener, Philip Levine, and R. E. Stetson—identified the Rh factor as the cause of the majority of transfusion reactions. The Rh factor is a protein substance found in the red blood cells of 85 percent of the population (they are referred to as Rh positive). Fifteen percent of people lack this factor (termed Rh negative). If Rh negative is transfused into someone who is Rh positive, the outcome may be a serious, even a fatal, reaction. Once a reliable way to test for Rh negative or Rh positive was identified, another important piece of knowledge had been acquired.

THE ESTABLISHMENT OF BLOOD BANKS

Need always stimulates progress, and when it came to blood transfusions physicians faced a new dilemma. Now that they could successfully categorize blood and perform transfusions successfully, they needed to devise a way to be able to get the right type of blood to the location where it was needed, and the need generally was created under emergency circumstances. During World War I, an increasing number of soldiers were being injured but surviving. This increased the need for available blood (and its availability in several types) so scientists began working hard to find ways to preserve and transport blood.

A physician in Chicago, Illinois, was one of the first to report notable progress. Dr. Bernard Fantus had read of a Russian doctor who saved cadaver blood to be reused, and he came up with a way for donated blood to be classified, labeled, and stored until needed. Dr. Fantus was the first to use the very appropriate term *blood bank,* and he used it to describe the facility he instituted at Cook County Hospital in Chicago. By the end of 1947, blood banks became more common in major cities throughout the United States, and hospitals and organizations such as the Red Cross began to encourage the public to donate blood as a helpful act of goodwill.

During the 1980s and 1990s, medical facilities had not yet fully mastered the ability to screen donated blood for diseases like Hepatitis C and HIV (human immunodeficiency virus), but today the testing of donated blood is greatly improved, and the federal government strictly enforces blood-screening tests to lessen the likelihood of disease transmission through transfusion.

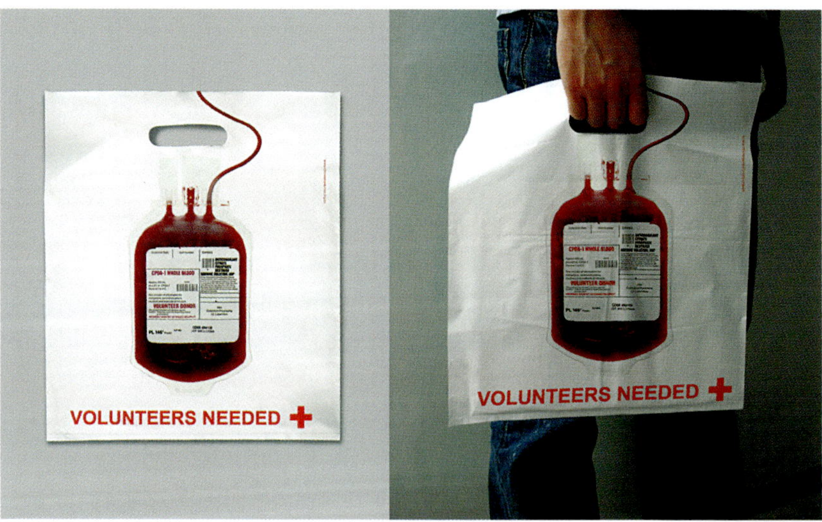

The blood bag was a significant development because it provides a safe and convenient way for blood to be transported. By replacing breakable glass bottles with rugged plastic bags, it was much safer and easier to transport blood. *(Red Cross)*

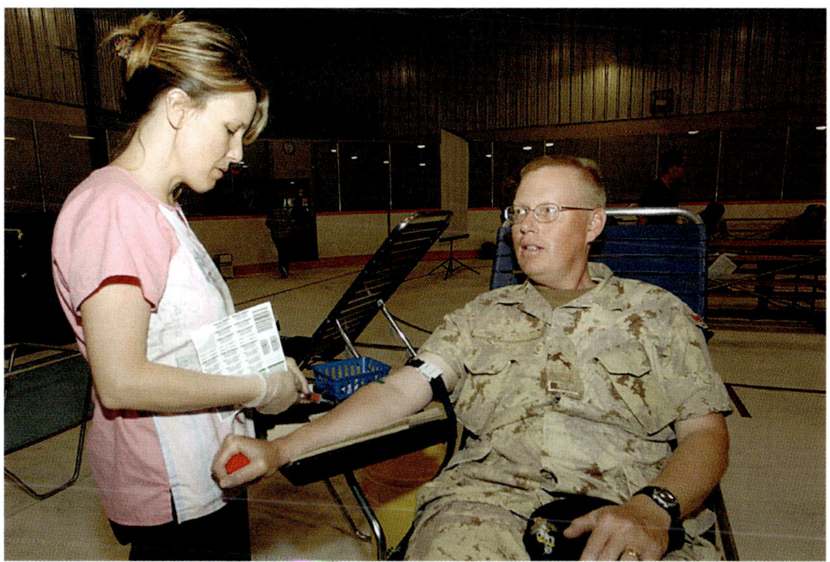

Drawing blood *(Master Corporal Dan Noiseux, Canadian Blood Services)*

CARDIAC SURGERY ADVANCES

During the 19th century, a few surgeons experimented with performing repairs to the sac that surrounds the heart, but the first successful surgery to the heart was performed in 1896 when a German surgeon successfully repaired a stab wound to a patient's right ventricle. As time went on, surgeons continued to contemplate how to conduct repairs to the heart, but they were well aware of the risks. As a result, not much progress was made until the situation was desperate during World War II. The U.S. Army surgeon Dr. Dwight Harken (1910–93) was stationed near the front lines in Europe. The patients who were coming into his hospital had shell fragments and bullets inside their hearts, and Harken was alarmed by what he saw. If these foreign elements were not removed from the heart, a cardiac infection or other damage was almost guaranteed, yet the patients would almost certainly die if surgery to remove the shrapnel from the heart was undertaken. Harken wanted to develop a surgical process that might work, so he began experimenting on animals, working to create a technique

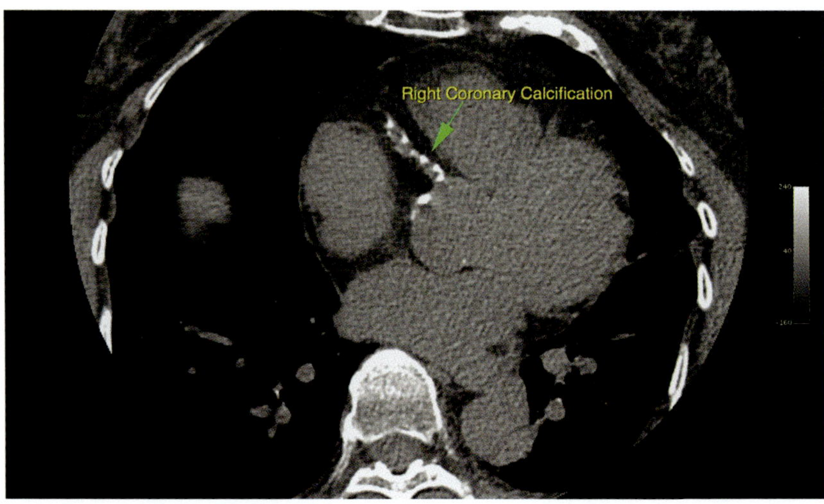

CT scan of heart *(Diagnostic and Wellness Center, Division of Cardiology at Harbor-UCLA)*

where he could open a small hole in the cardiac wall and insert his finger to remove the foreign element. His early results were poor, but he began to see some successes with practice; animals were surviving the surgery. By the time he felt confident enough to try the surgery on soldiers, most procedures went quite well. The significance of this surgery exceeded the lives saved, because for the first time, a surgeon had proven that it was possible to operate on the heart.

Despite Harken's success, both physicians and surgeons were aware that the variety of problems that could affect the heart were numerous, and they still lacked a blueprint for other types of heart surgery. They did not know what to do about children born with congenital heart defects or victims whose heart valves were narrowed or stuck, so they knew that additional experimentation was necessary. Surgeons needed to be able to work inside the heart without their patients bleeding to death. They had learned that a patient's circulation could be stopped temporarily permitting some surgery, but they found that this technique provided surgeons with only about four minutes before the brain became damaged from oxygen deprivation. Over time, scientists and physicians developed increasingly sophisticated ways to perform heart surgery, as follows:

- Cooling the body. As medical professionals looked for ways to effectively operate on the heart, a Canadian surgeon named Dr. Bill Bigelow who worked at the University of Minnesota was developing a theory. He had noted that when animals hibernated, their hearts beat more slowly, which permitted them to survive for months without food. Bigelow began animal experiments and found that when dogs' bodies were cooled, it provided a surgeon with a longer time span in which to operate, and the dogs didn't die. Bigelow showed that at lower temperatures, the tissues of the body and brain did not need as much oxygen and could survive without oxygenated blood for longer. In 1952, Bigelow was encouraged to test his method on a child who was desperate for heart surgery. Two surgeons at the University of Minnesota were able to operate successfully on a five-year-old with a heart defect by creating a state of hypothermia. The patient's body temperature was brought down by covering her with a cooling blanket. This permitted the surgeons to clamp off the inflow to the heart so they could make the repair. They estimated they had only 10 minutes in which to work, so they operated quickly and then immersed the patient in warm water. The surgery was deemed a success. Though this process became useful for treating relatively minor problems, surgeons continued to look for ways that would extend their operating time.

- Maintaining circulation by perfecting open heart surgery. Surgeons needed a way in which the heart could be worked on while maintaining circulation. After experimenting with artificial pumps outside the body, in 1954 Dr. C. Walton Lillehei (1918–99), one of the physicians who had mastered the 10-minute technique, established a technique in which the patient's mother or father was used as a heart-lung machine. The first patient was an 11-year-old boy with a ventricular septal defect. The boy's father was anesthetized next to his son, and the boy's blood was routed through the father's system where it could be oxygenated before returning to the boy's body via the carotid artery. In this particular case, the method provided the surgeons with 19 minutes to make the repair. Lillehei is often referred to as the father of open heart surgery, since he led the way.

- In 1931, a physician named John Gibbon (1903–73) was greatly affected by the loss of a young patient to heart disease. He determined that there had to be a way to create a heart-lung bypass machine, and he began experimenting with various devices, using animals for the test cases. In 1935, Gibbon created a machine that successfully kept a cat alive for 26 minutes. His work was interrupted by the necessity of military service. After World War II, Gibbon returned to work on his device, and he experimented with his method on a dozen dogs; some he kept alive for up to an hour. In 1953, Cecelia Bavolek was the first human on which the heart-lung bypass device was used. The surgery was a success, and by 1960 use of the device was common. Since the 1990s, surgeons have begun to perform coronary artery bypass surgery without the exterior cardiopulmonary bypass. In these operations, the heart is beating during surgery, but it is stabilized to provide an almost still work area. Some researchers believe this approach results in fewer postoperative complications.

- Minimally invasive surgery. A new form of heart surgery that has grown in popularity is robot-assisted heart surgery. The surgery is performed by a robotic machine controlled by the heart surgeon and, because the robotic hands can be quite delicate, procedures can be performed with much smaller incisions. This makes the surgical process easier for the patient, and the recovery time is greatly reduced.

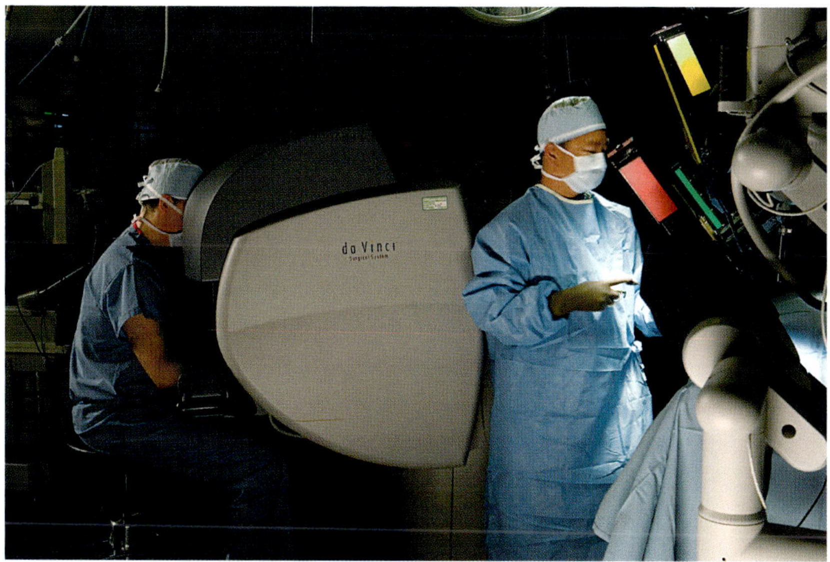

The da Vinci robotic system is intended to be minimally invasive. *(Global Robotics Institute, Florida Hospital)*

Progress continues, and during the 21st century there will undoubtedly be even more advanced ways of performing heart surgery.

ARTIFICIAL HEARTS AND TRANSPLANTS

As scientists came to understand that the primary job of the heart is purely mechanical—that it works as a pump to keep blood circulating through the body—they began to explore whether there could be a way to replace it. They contemplated using an artificial heart or even another human heart, and they experimented with both. As early as 1935, a French surgeon named Alexis Carrel (1873–1944) working with American aviator Charles Lindbergh (1902–74) designed a heat pump that was intended to work outside the body to keep blood circulating through the organs while a surgeon worked to repair the person's heart.

As with other transplant experiments, an animal was the first recipient of a fully functional artificial heart (called a TAH). In 1957 at a clinic in Cleveland, Ohio, two surgeons Willem Kolff and T. Akutsu implanted an artificial heart in a dog. Kolff dedicated his career to working to perfect the artificial heart, and he was rewarded by others taking his work seriously. The government took serious interest in these advances, and in 1964 the National Institutes of Health established an artificial heart program to encourage more research on these devices.

One of the pioneers of cardiac surgery Michael DeBakey (1908–2008) advanced medical science by his work on heart surgery, and in 1966 he also designed and implanted a left ventricular assist device (LVAD). This mechanism worked via air pressure and could be used to help pump blood out of the heart and into the arteries. (DeBakey is also mentioned in chapter 6 for developing a plan for MASH units to provide better medical care on the battlefield.) Serious heart disease generally involves failure of the left ventricle, so DeBakey's invention moved the field forward in

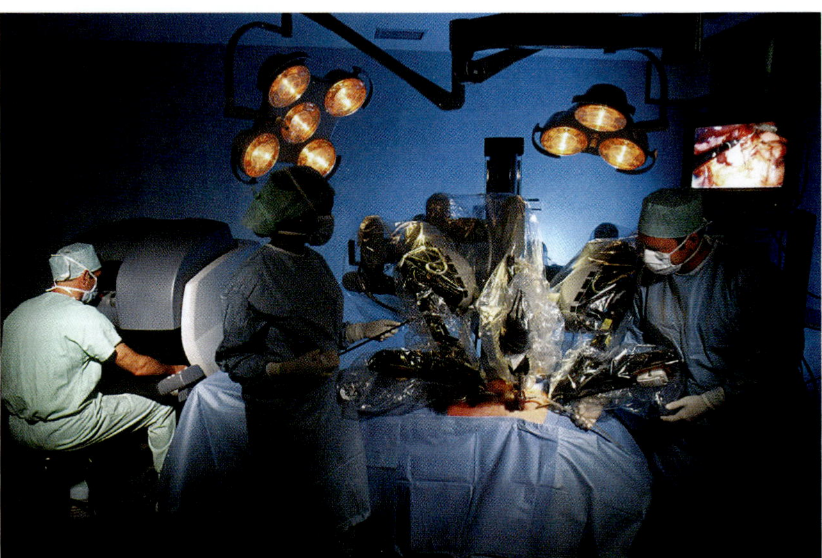

The U.S. military initiated research on operating robots as a method of allowing experienced trauma surgeons to operate on battlefield casualties from a remote location. Recent technological advances have allowed incorporation of robotics into routine clinical care. *(Loma Linda University Medical Center)*

an important way. (See the sidebar "Michael DeBakey, Doctor and Patient" on page 131.)

Human Heart Transplants Advance

The next step forward in heart replacement was made by a surgeon in South Africa, Christiaan Barnard (1922–2001), who had been experimenting with heart transplants using dogs as subjects. By the end of 1967, Barnard felt prepared to try his techniques on a human subject and was continuing to practice his methods until the right circumstances presented themselves. Among the possible candidates for a heart transplant was 55-year-old Louis Washkansky, a former athlete suffering from bad health. He was diabetic, had experienced three heart attacks, and was suffering from congestive heart failure. When Dr. Barnard was made aware that a young woman with a functional heart had died from injuries sustained in an auto accident, Barnard decided it was the right opportunity to attempt a heart transplant and he placed her heart in Washkansky. Louis Washkansky did quite well immediately after surgery, but he died of pneumonia 18 days later. His immune system was so weakened from the drugs and radiation to prevent the body's rejection of the new organ that he could not fight off infection.

Barnard's heart transplant surgery was initially celebrated when the patient seemed to be doing well, and then after Washkansky died the questions poured in. Scientists, physicians, and the public all voiced their opinions about whether or not this type of surgery should be done. Christiaan Barnard was seemingly unmoved by the concern, and he kept on practicing the procedure whenever circumstances presented themselves. By 1983, more than 63 heart transplants had been done under Barnard's direction at the hospital in Cape Town.

One of the problems with any type of transplant, including heart transplants, has to do with the body's natural instinct to reject the new organ. The American surgeon Dr. Norman Shumway (1923–2006) created a team of scientists and physicians to puzzle through the complex biological problem of organ rejection. The team devised a way to monitor the heart (via a catheter) for

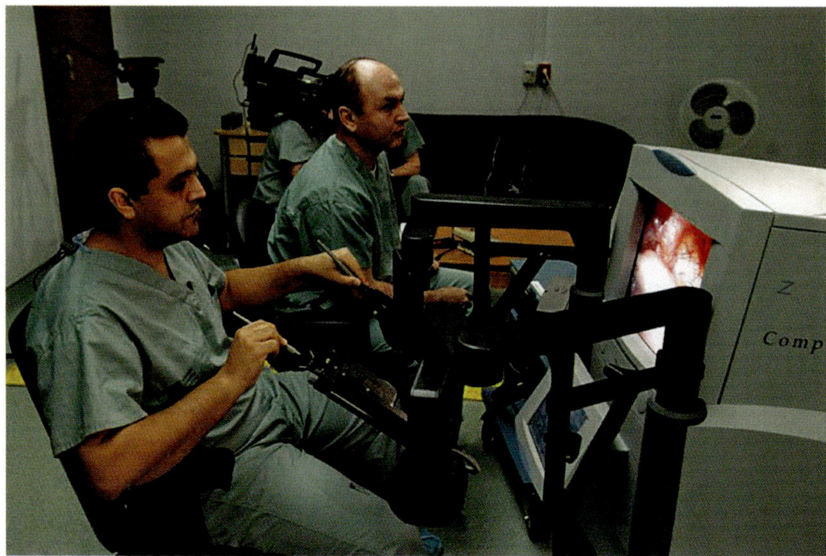

The concept of using robots in surgery is attractive because it combines the precision and accuracy of a machine with the judgment of an experienced surgeon.

signs of rejection so that the higher doses of the immunosuppressive drugs were increased only as needed. In the meantime, scientists in Norway identified a fungus that revolutionized transplant surgery. The substance was cyclosporin, and it appeared to have the perfect immunosuppressant properties—controlling organ rejection without knocking out all resistance to infection. In 1980, Dr. Shumway's team was the first to embrace the new medicine, and it transformed the picture for heart transplant recipients.

Transplant surgeons today face a new problem: finding enough healthy hearts. In the United States alone, 2 million people suffer from congestive heart failure. When drug treatments fail, transplants are the best hope. But fewer than 2,500 donor hearts are available each year.

Later Progress in Artificial Hearts

Artificial hearts were originally intended as a temporary solution. A mechanical heart would be introduced to keep a person living until an available heart was found. The method was first attempted in 1969, and a dozen years later doctors were beginning to wonder how much time could be bought by the use of an artificial heart. In 1982, the dentist Barney Clark became famous for having received an artificial heart, the Jarvik-7, the first heart that

MICHAEL DEBAKEY, DOCTOR AND PATIENT

Michael Ellis DeBakey was a noted physician and surgeon whose work in cardiovascular surgery will long be remembered. He created numerous procedures and devices that were very helpful to the field of heart health. At the age of 97 (in 2005), Dr. DeBakey suffered a tear in the wall of his aorta. Ironically, DeBakey had created a procedure to treat exactly this type of condition. He still was a member of a cardiovascular medical practice in Houston, Texas, and his partners wanted to perform the DeBakey procedure, as it is known, on patient DeBakey. DeBakey was alert enough to be aware of the surgeons' desires, but he asked that they not do it—he felt he was too old for the surgery.

DeBakey's condition deteriorated, and when he became unresponsive the surgical team opted to proceed. Because they did not have his consent, the Houston Methodist Hospital ethics committee had to be consulted. Though the decision was far from unanimous, the surgeons were given permission to proceed. In a seven-hour surgery, DeBakey became the oldest patient ever to undergo the surgery he had created. His postoperative recovery was difficult, and DeBakey remained in the hospital for the next eight months at a cost of more than $1 million. In September 2006, Dr. DeBakey was released from the hospital, his health returned, and he went back to a limited schedule at work. He died of natural causes in 2008 at the age of 99.

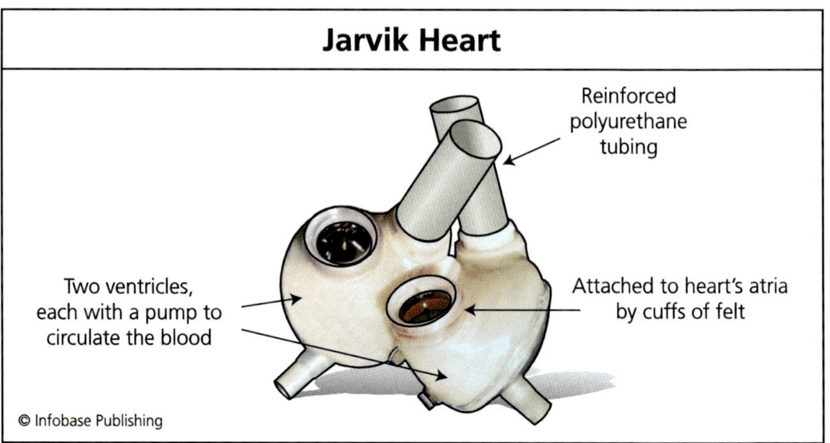

Scientists were taken with the idea that it might be possible to create an artificial heart that could do the job of the human heart. Robert Jarvik created one of the first.

was intended to be a permanent solution. In addition to difficult surgery, there were still technological problems to overcome. The Jarvik-7 was powered by compressed air, and at that time the only way to provide the air was via an air compressor outside the body, so the patient had to be surgically connected to an external unit at all times. Clark survived for 112 days.

The surgery team had been led by William DeVries who went on to implant other Jarvik-7 devices. One of his next patients, William Schroeder, survived 620 days, but he suffered many setbacks. Physicians as well as the public realized there was still a lot to learn about this process.

As science has progressed, smaller has been deemed better, and in late 2008, French researchers developed an artificial heart made of biosynthetic tissues that are chemically treated to prevent human immune systems from rejecting the heart. The heart itself reproduces the physiology of a normal heart, and its beat is powered by batteries that are intended to last five years. The device has not yet gone through clinical trials, but scientists are hoping that it will be available as an alternative to a transplant by 2013.

PLASMA THERAPY:
A Possible New Sports Treatment

In a giant step away from bloodletting, the latest form of therapy being used for some sports injuries involves reinjecting a patient with his or her own blood that has been processed to enrich its platelets. (Platelets are involved in releasing proteins and other particles that encourage the body's self-healing.)

The method involves injecting portions of a patient's blood directly into the injured area. This seems to encourage the body's natural effort to repair muscle, bone, and tissue. The process is not considered difficult by physicians and has been tested on professional athletes whose injuries keep them off the field, the court, or the baseball diamond. According to the *New York Times* (February 16, 2009), the method was used on Takashi Saito, a pitcher for the Los Angeles Dodgers who suffered a partially torn ligament in his throwing arm. By injecting Saito's elbow with his own platelet-enriched blood, he was able to return to pitching within a couple of months. Surgery would have sidelined him for about a year. While the team doctor noted that 25 percent of these injuries heal on their own (and certainly there is no way to know if Saito's would have), medical professionals are encouraged by what they are seeing, and clinical trials are underway.

The process of creating blood plasma with a higher platelet count involves withdrawing a small amount of the patient's blood and putting it in a filtration system that separates the platelets from red blood cells. The physician then injects a very small quantity (a teaspoon or two) of the substance into the area where the person has been injured. The platelet-rich plasma seems to enhance the body's ability to

(continues)

(continued)

grow new tissue or bone cells without causing problems with clotting.

While those who manage or invest in professional sports teams are very excited about healing their players quickly, other people speculate that there will also be a market for this procedure among weekend warriors—recreational athletes who love sports and hate missing a weekend of tennis, golf, or handball. The procedure could be done in much less time and at a lower cost than surgery, which is often required with chronic sports injuries.

CURRENT THINKING ON HEART HEALTH

According to the American Heart Association (AHA), about 60 million Americans currently suffer one cardiovascular condition or another, and almost 1 million deaths in the United States are attributed to cardiovascular disease each year. Heart disease is the leading cause of death for both men and women in the United States. With explosive increases in obesity and type 2 diabetes and with the baby-boom generation aging, the problem of cardiovascular disease shows no signs of going away.

Today, doctors know that lifestyle can make a big difference in controlling the risk factors of heart disease, which are usually signaled by high cholesterol, high blood pressure, or both. Rather than having patients reach a critical point where they need an artificial heart or a heart transplant, doctors today—along with the federal government—are trying to reduce the deaths from heart disease by encouraging healthy lifestyles. To improve heart health, today's medical practitioners encourage people to eat a healthy diet, maintain a recommended weight for their build, exercise regularly, quit smoking (or not start), and minimize stress—all factors that can lead to heart disease.

CONCLUSION

People living only 150 years ago might well have undergone bloodletting as part of a medical treatment, so the fact that today blood transfusions, open heart surgery, and heart transplants are conducted with frequency is all the more amazing. While physicians would prefer to teach people to live healthy lifestyles so that heart transplants and heart surgery are unnecessary, the reality is that these medical treatments are going to continue to be needed. Advances in fixing a patient's own heart will be a top priority, but with the difficulty in finding hearts available to transplant, the most promising frontier may be in the creation of miniature artificial hearts.

8

DNA Changes the Medical Knowledge Base

Increasingly, the press is writing about amazing progress in science and in medicine because of the relatively newfound ability to identify a person's DNA. Since the late 1950s and early 1960s, molecular biologists have learned to characterize, isolate, and manipulate the molecular components of cells and organisms. These components include DNA (deoxyribonucleic acid), the repository of genetic information; RNA, a close relative of DNA, whose functions range from serving as a temporary working copy of DNA to actual structural and enzymatic functions; and proteins, the major structural and enzymatic molecules in cells. In molecular biology, scientists study how various cell systems interact as well as the interrelationships involved in DNA, RNA, and protein synthesis. It is a field where the study of biology, chemistry, genetics, and biochemistry overlap.

DNA has to do with the molecular makeup of every living thing, and scientists believe that the future of medicine lies in individually tailoring cures for each person or each disease, or even both. Every cell (except red blood cells) within each human or living thing contains a copy of the same DNA. The DNA sequence is the particular side-by-side arrangements of bases along the DNA strand, and this order spells out the exact instructions required to provide unique traits for each particular organism as well as how it develops and functions. As scientists come to better understand the DNA of each human being as well as the DNA of various illnesses, they feel there will be notable medical breakthroughs.

Before DNA could even be conceived of, advances in understanding genetics and inherited traits were necessary. Ancient people understood some basics about breeding and inheritance in animals: Certain animals, when bred together, begat animals with the same strengths as the parent animal. While some cultures worked to refine these planned breedings, it was mostly a hit-or-miss process, and no one had any idea about the mechanics of how this worked scientifically.

When the structure of DNA was discovered by James Watson and Francis Crick, it was referred to as the secret of life, but today scientists know DNA is not really the secret to life. It is more like a special key—with it, they have only just begun to unlock some of the mysteries of human development, which in turn, may make possible a new form of personalized medicine. This chapter explains how DNA was discovered and how it may affect medical developments of the future. Most people know of DNA's application in crime cases, and the chapter also explains how DNA helps solve crimes. (See the sidebar "DNA and the Criminal Justice System" on page 142.)

THE BASICS OF DNA AND HOW GENETICS WERE VIEWED IN THE PAST

Before genetics could become a science, the world needed to discover how life began scientifically, and they needed to know that there were mechanisms—genes—in the human body that somewhat reliably determined inheritance. The work of the Austrian monk Gregor Mendel (1822–84) in the mid-19th century would eventually be very helpful to scientists, but it was a long time before anyone knew what he had learned.

Crick and Watson with their DNA model *(A. Barrington Brown/Science Photo Library)*

Mendel was an amateur gardener who enjoyed crossbreeding pea plants. Mendel kept careful records of his crossbreeding of tall or short plants that had smooth or wrinkled peas and began to see that something within the plant-breeding process operated with an orderly, dominant/recessive plan. (When he bred tall plants with short ones, he did not get medium-sized ones; he always got tall ones.) There was not much of a way to communicate for a monk with an interesting hobby, so no one knew of his contributions during his lifetime. But he left behind meticulous records and eventually his work was found by others, and it prepared the way for some very significant discoveries. Finally in 1900, two scientists working separately on the study of genetic traits came upon the work of the then-deceased Gregor Mendel. When Hugo de Vries and Carl Correns released to the world Mendel's 1865 research, it marked the beginning of the study of modern genetics.

The next pertinent discovery was made in 1944 when Oswald Avery (1877–1955), an American scientist and early molecular biologist, proved that DNA carries genetic information. While this discovery fascinated scientists, there was still little they could do with the information because they did not yet know the molecular structure of DNA. Only by decoding it could they begin to understand it—and understand how to use that information to help humankind.

SCIENTISTS AND SERENDIPITY

While almost all scientific discoveries rely heavily on the scientific work that preceded it, the discovery of the structure of DNA relied not only on the work of others, but also a great deal on serendipity. In addition, there were several scientists (including teams of scientists) working competitively to successfully be the first to identify the structure of DNA. Watson and Crick's victory was due to a little luck and a lot of careful attention paid to their own work as well as what they were hearing from others. Added to this mix was the fact that one of the front runner scientists (Rosalind Franklin) was abrasive and therefore not well liked, which encouraged members of her department to go against protocol and share with others some of her findings.

The British biophysicist Maurice Wilkins (1916–2004) was among the first who were working to determine the structure of the DNA molecule. Wilkins, who was based at King's College in London, needed an image of it, so he turned to Rosalind Franklin (1920–58), who was only a graduate student but was acknowledged to be the best person at performing X-ray crystallography. Franklin prepared the necessary images and in 1951 agreed to give a departmental talk about her X-rays of a dry and a wet form of DNA, which were beginning to provide evidence of the helical structure of DNA. As it happened, James Dewey Watson (1928–   ), a 23-year-old American zoology graduate who had studied ornithology and then viruses, visited the department that day and sat in on Franklin's lecture, which gave him some important clues about Franklin's early findings.

Watson had only recently arrived at Cavendish Laboratory, where he hoped to study DNA. He had just become interested in the subject, but he had already decided that it was his goal to be the one who decoded the structure. At Cavendish, he encountered his future partner Francis Crick (1916–2004), a physicist who had become fascinated by the application of physics to biology and who was also intent on investigating how genetic information might be stored in molecular form.

Though an odd pair, Watson and Crick were united by their determination to better understand DNA. Crick brought knowledge of X-ray diffraction, and Watson brought knowledge of viruses and bacterial genetics. Watson and Crick were actually assigned to work on another type of project, but they secretly kept working on the unraveling of DNA. They respected Franklin's work and invited her to come meet with them but she turned them down.

On another continent, the American Nobel Prize–winning chemist Linus Pauling (1901–94) was working with X-ray crystallography and molecular model building. In 1952, he hoped to travel to England to a conference he knew Franklin would be attending, but he was denied a passport because of allegations that he was a Communist sympathizer.

In the meantime, Francis Crick was annoyed that he and Watson were not officially permitted to work on DNA. Crick made another appeal to their superior, and permission was finally granted.

Franklin continued to make good progress with her work. She had successfully developed a good photograph of the B (wet) form of DNA, which showed a double helix. However, she was not ready to release her information until she had further explored something that was bothering her about the A (dry) form. Her reluctance annoyed her partner, Wilkins, who decided to move on without her. In 1953, Watson dropped in to visit Wilkins, and during the course of their meeting Wilkins happened to show Watson a copy of Franklin's photograph of the wet form of DNA, revealing the helical form that Watson suspected. The photograph led Watson to suspect that DNA could reproduce because it was structured as a double helix.

Later on, without Franklin's permission, a departmental report of Franklin's was passed to Watson and Crick. It offered conclusive evidence that DNA was a multiple helix, and Watson and Crick learned that the phosphate backbones of DNA should be on the outside of the molecule. This finding was key to figuring out the structure. (It will long be debated whether Watson and Crick should have had access to Franklin's results before she formally published them herself.)

Watson and Crick had not found the answer, but they knew they were close. They turned to work done by the biochemist Erwin Chargaff in 1950. Once again, Franklin was the scientist who really had a full understanding of Chargaff's work, which involved the base pairings of DNA. She had completed a draft of a paper, dated March 17, 1953, that identified the double-helix structure of DNA as well as the specific base pairings that permit the unzipping of the double helix during cell division so that the gene, and eventually the chromosome, can replicate.

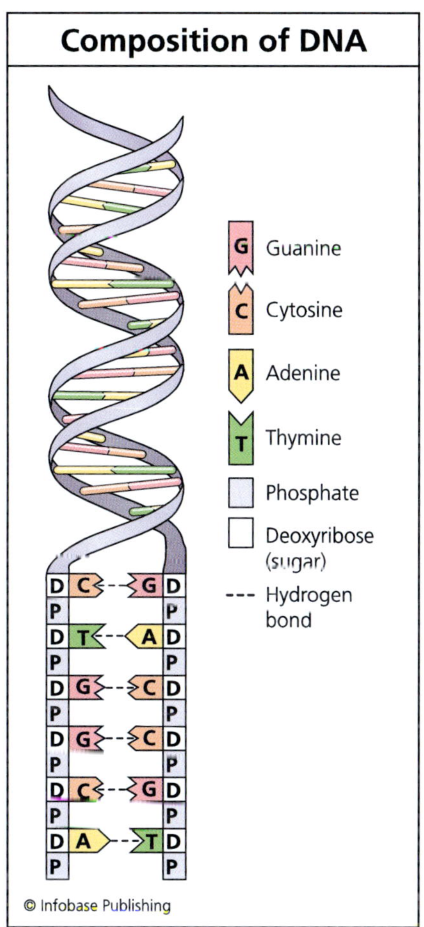

Watson and Crick were able to decode DNA to determine how the genetic code is carried.

Despite Franklin's overall understanding of the process, it was Watson and Crick who got to press first with their paper. Without much of a fuss, Watson and Crick's paper on the structure of DNA appeared in the British journal *Nature,* and it described the DNA molecule as a long, two-stranded chain coiled into a double helix and resembling a twisted ladder. Their paper summed up the contribution of Wilkins and Franklin by simply mentioning that their thoughts were "stimulated" by the unpublished results of Wilkins, Franklin, and their coworkers at King's College.

In 1962, Watson, Crick, and Maurice Wilkins were given the Nobel Prize in physiology or medicine for their work. Franklin had died in 1958 from cancer, possibly related to her extensive expo-

DNA AND THE CRIMINAL JUSTICE SYSTEM

Though the sequencing of DNA has now been around since the 1950s, its use by the criminal justice system is a relatively new development despite how common it is as a part of today's television crime shows. The National Academy of Sciences approved the use of DNA for court cases in 1992.

The first attempt to use DNA in a criminal case was in Britain in 1986 when Professor Alec Jeffreys assisted in solving a pair of rape-murders of two teenagers. When two 15-year-old girls were murdered in 1983 and 1986, police originally arrested a young man with a history of mental illness. Professor Jeffreys was not convinced that the police had the right man. After analyzing semen collected from both bodies, Jeffreys asked everyone in the small town of Narborough to voluntarily submit to DNA testing. At first, no guilty party emerged. Eventually, police found that a baker by the name of Colin Pitchfork had paid someone else to give a blood sample for him. Police reinterviewed Pitchfork, and he confessed to both crimes. When he provided his own DNA, the police had a match and their man.

sure to radiation in her work with the X-ray diffraction that was so vital to better understanding the structure of DNA. Because the Nobel cannot be given posthumously, she was never honored for her work.

THE HUMAN GENOME PROJECT

By the 1980s, James Watson had another passion to pursue, and this too has been absolutely key to medical and scientific advances. Watson helped lobby Congress to create the U.S. Human Genome Project, the multimillion-dollar effort to map out the exact *nucleotide* sequence contained in each of the 24

An early test of DNA in the courtroom in the United States occurred in what has become known as the trial of the century—the 1995 criminal murder trial of sports star O. J. Simpson. Though Simpson's DNA was found at the crime scene, the prosecution was unable to convince the jury of Simpson's guilt. The use of DNA in court was still very new, and though scientists and lawyers faced off on the issue, the defense successfully convinced the jury that the evidence may have been contaminated. This case brought to light the importance of careful training of criminal investigators in the collection of evidence, and as a result crime laboratories realized the importance of being certified for this type of testing.

By the late 1990s, forensics labs started to adopt a new method of analysis called STR (short tandem repeats) that cuts analysis time from weeks to days and uses patterns that repeat just a few times, between five and 30 in most cases. And it also improves accuracy—early on in DNA forensics the chance of error was one in 100,000. The STR method makes it more like one in 1 trillion.

human chromosomes—the so-called book of life (consisting of approximately 3 billion letters).

Ironically, the Human Genome Project grew out of the U.S. Department of Energy (DOE) (albeit, it did originate in its health and environmental program). Since 1947, the DOE and its predecessor agencies have been charged by Congress with developing new energy sources and pursuing a deeper understanding of potential health and environmental risks posed by their production and use.

In 1986, Charles DeLisi, who was then director of DOE's health-related research programs, became convinced that if they were going to be able to effectively study the biological effects of radiation (along with information on whether these effects were passed on genetically, such as in the cases of survivors of Hiroshima), he needed a way to do so quickly. By 1990, DOE and the National Institutes of Health, who understood that knowledge of the human genome was necessary to the continuing progress of medicine and other health sciences, had agreed to devote $3 billion to the project, and they established a 15-year time line. They were joined by the Wellcome Trust, a private charitable organization in Britain. There were also contributions from Japan, France, Germany, and China.

In 1998, a private firm named Celera Genomics, run by researcher Craig Venter, became involved in this type of research. Venter was using a newer technique (shotgun sequencing, a process which breaks the DNA into shorter segments that permits faster readings; the segments are rejoined for a complete reading at the end) and hoped to finish before the government and patent some of what he found. Though his contributions were notable, in March 2000 President Clinton announced that the genome sequence should not be patented, sending biotech stocks—including Celera's—plummeting.

The competition proved healthy, however, and by 2000, due to widespread cooperation, both public and private, a rough draft of the genome was actually finished early and was jointly announced by then-president Clinton and British prime minister Tony Blair.

In February 2001, both Celera and the government scientists published details of their discoveries—*Nature* published the government's version, and *Science* published Celera's. Together, the sequence they had identified made up about 90 percent of the genome. By 2003, a joint release announced that 99 percent of the genome has been sequenced with 99.99 percent accuracy. For all ostensible purposes, the project was completed in April 2003, bringing it in two years ahead of schedule. This happened to be the 50th anniversary of Watson and Crick's publication of DNA's structure that launched the era of molecular biology.

What has surprised everyone—scientists included—is that humans are a great deal simpler than what was originally imagined. They have found that the human genome has only about 30,000 genes; the original estimate had been three times that number. This finding suggests that a great deal more has to be learned about how genes function and how their instructions are carried out as well as how they produce diseases and other anomalies. But in the meantime, the gains have been incredible.

The Human Genome: What It Is

The human genome is a person's complete set of DNA, arranged into 23 distinct chromosome pairs (the 24th pair is the set that determines gender)—physically separate molecules that range in length from about 50 million to 250 million base pairs. Each chromosome contains many genes, the basic physical and functional units of heredity.

Genes make up only about 2 percent of the human genome; the remainder consists of noncoding regions, whose functions may include providing chromosomal structural integrity and regulating where, when, and in what quantity proteins are made. Although genes get a lot of attention, it is the proteins they make that perform most life functions and even make up the majority of cellular structure.

DNA AND THE FUTURE

Knowledge about DNA may lead to understanding about how all the parts of cells—genes, proteins, and many other molecules—

work together to create complex living organisms. DNA underlies almost every aspect of human health, and understanding about what DNA has to do with health will have a profound impact

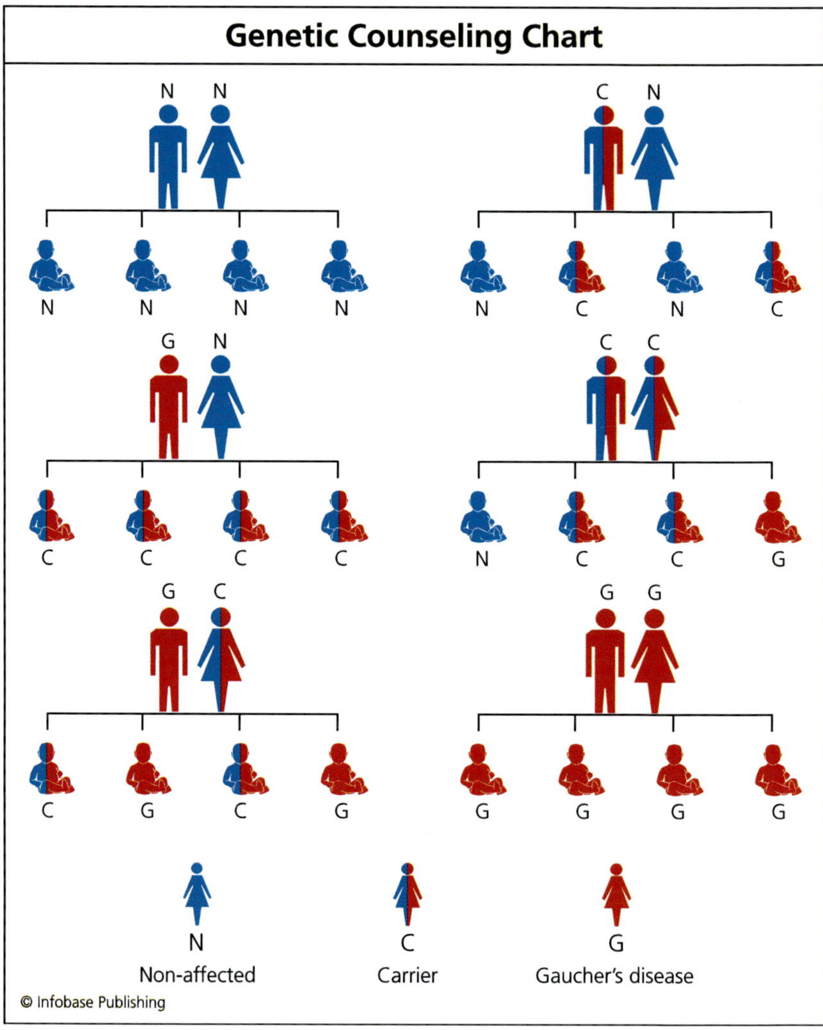

Genetic counseling has traditionally been most useful for single gene disorders like Huntington's disease or Tay-Sachs disease, for which there are genetic diagnostic tests. For such diseases, affected patients may wonder about the risks of passing on an illness to their children, while children of affected patients will want to know their statistical risk of developing the disease.

on the ways disorders are diagnosed, treated, and prevented. For example, scientists have discovered a variant gene carried by more than one-third of Americans that leads to a sizable increase in the risk of type 2 diabetes. This should lead to an improvement in diagnostic testing as well as in treatment.

In December 2005, a new project, the Cancer Genome Atlas, was announced. The Cancer Genome Atlas Pilot Project is designed to identify and unlock the genetic abnormalities that contribute to cancer—an effort that could lead to new diagnostic tests and treatments for the disease. Scientists have long known that genetic mutations accumulate in a person's normal cells over a lifetime and can make those cells cancerous. About 300 genes involved in cancer are already known, and there are a handful of drugs that work by interfering with specific genetic abnormalities.

All this information about the human genome is now available to researchers worldwide, and a brand-new future is opening for the human genome reference sequence. Among the changes that may occur are these:

- Exploration of the evolutionary process. Scientists are comparing the lineage of chromosomes in various species to better understand what changes there have been in various organisms and why they occurred.

- Gene testing. Companies now offer inexpensive and easy to-administer genetic tests that can show a predisposition to a variety of illnesses such as breast cancer, blood clotting, cystic fibrosis, liver diseases, and so on. Experts are greeting these tests with mixed reactions. A predisposition to an illness does not guarantee an illness, and society must be careful where this information is stored and how it is used.

- Gene therapy enhancement. One day one's own tissues could help replace cells damaged by injury or disease. DNA could be withdrawn and used to prime embryonic cells that are ready to serve as replacement cells to bolster normal function or to create immunity.

- Pharmacogenomics. This type of medicine will lead away from one-size-fits-all medicine. In the future, both treatments and vaccines will be made to order for each individual. This will offer direct lifesaving benefits, as today more than 100,000 people die each year from bad reactions to medications. DNA will help predict what will work and for whom.
- Improved vaccines. In 2009, the government announced a new swine flu vaccine, and scientists think they will be able to identify the exact flu strain and manufacture it quickly with the help of genetic engineering.
- Genetically modified foods. Scientists are already beginning to create healthier crops based on understanding the genetics of different plants.

As scientists continue to pursue these leads, this list of undertakings will certainly change and grow.

ETHICAL ISSUES REQUIRE SOLUTIONS

Right alongside the scientists making all the genetic advances, scientific committees will need to wrestle with the ethical issues that these advances raise. These can range from the seemingly simple (what happens if genetically altered corn cross-pollinates with regular corn?) to issues that will affect people in larger ways. For example, if someone undergoes genetic testing that shows a predisposition to some type of chronic illness, scientists have to be certain that this can never be used to discriminate against someone. Predisposition to a disease does not mean a person will get the disease. It means that if the correct influences occur, a person might possibly develop the disease. For example, the person who has a predisposition to lung cancer might be more likely to develop it than someone else after exposure to secondhand smoke. This would spark the nature versus nurture debate.

Questions about privacy, fairness in use, and access to genetic information all need to be answered as advances continue.

NATIONAL INSTITUTES OF HEALTH: UNDIAGNOSED CASES

In late 2008, the National Institutes of Health (NIH) announced a new clinical research program that aims to provide answers to patients with mysterious conditions that have long eluded diagnosis. The effort of this special unit will be to use biomedical research to make molecular diagnoses and examine the disease through the genes involved. The goal will be to unravel these cases both to advise the individual but also to advance medical knowledge in general.

To evaluate each patient enrolled in the new program, the NIH will enlist the expertise of more than 25 of its senior attending physicians, whose specialties include endocrinology, immunology, oncology, dermatology, dentistry, cardiology, and genetics. Though the patients accepted into the program will spend a week on site being evaluated by the various experts, the work will continue long after the patient goes home. Blood and tissue samples for the various conditions will be carefully scrutinized to piece together genetic information with the intent that some helpful answers will be available to the patient and to future patients who exhibit similar symptoms.

CONCLUSION

When scientists and physicians contemplate the personalized medicine that will eventually be made possible because of research into DNA, medical science is only at the beginning of the journey. Over the course of the next few years, scientists will be devising more and better ways to treat each person based on the person's own genetics as well as the genetics of their particular illness.

1840s — Dorothea Dix begins campaign to improve care of the insane.

1842 — First surgical operation is performed using anesthesia.

1847 — Ignaz Semmelweis discovers how to prevent puerperal fever.

1849 — Elizabeth Blackwell is first woman to gain a medical degree.

1854 — Florence Nightingale and 38 other nurses go to the Crimea.

1855 — The nurse Mary Seacole establishes a British hotel in Crimea.

1858 — Rudolf Virchow proposes cell theory.

1862 — Pasteur refines what becomes known as pasteurization.

1864 — Pasteur is recognized for experiments that debunk spontaneous generation.

1867 — Joseph Lister publishes *Antiseptic Principle of the Practice of Surgery*, partially based on Pasteur's work.

1870 — Louis Pasteur and Robert Koch establish germ theory of disease.

1873 — Linda Richards becomes America's first trained nurse after graduating from the New England Hospital Training School.

1881 — Clara Barton founds the American Red Cross.

1881–82 — Pasteur develops anthrax and rabies vaccines.

1883 — Robert Koch issues his three postulates that explain the causes of disease.

1888 — The Pasteur Institute is founded.

1890	Emil von Behring discovers antitoxins and uses them to develop tetanus and diphtheria vaccines.
1890s	Viruses are identified (but not seen until electron microscope invented in 1930).
1894	The first known polio epidemic in the United States occurs in Vermont.
1895	The German physicist Wilhelm Conrad Röntgen discovers X-rays.
1899–1901	Walter Reed heads a commission that finally determines that yellow fever is spread by mosquitoes; this provides a way to diminish the contagion.
1901	Karl Landsteiner discovers different human blood types.
1903	Marie and Pierre Curie and Antoine Becquerel are awarded the Nobel Prize in physics for their work on radioactivity.
1906	The United States passes the Pure Food and Drug Act.
1911	Marie Curie is awarded the Nobel Prize in chemistry for discovering radium and polonium and isolating radium.
	Harvard creates department of industrial medicine for Alice Hamilton, the first woman to be hired as a faculty member.
	The German researcher Paul Ehrlich tests Salvarsan, first treatment effective against syphilis; regarded as birth of modern chemotherapy.
1915	Bayer introduces aspirin in tablet form.
1920s	Vaccines created for diphtheria, pertussis, tuberculosis, tetanus.
1921	Franklin D. Roosevelt is diagnosed with polio.
1928	Alexander Fleming discovers penicillin.

1937	The first blood bank is established in the United States.
1944	The molecular biologist Oswald Avery proves that DNA carries genetic information.
1952	Worst U.S. polio epidemic with almost 58,000 cases; Jonas Salk develops first polio vaccine and starts trials.
1953	Watson and Crick identify the structure of DNA.
1957	First artificial heart tested in a dog
1960	Oral birth control pill approved by Food and Drug Administration
1961	Albert Sabin develops oral vaccine with live virus.
1967	First heart transplant
1969	First artificial heart implanted in a human
1981	First outbreak of what is later realized to be AIDS
2003	99.99 percent of the human genome sequenced
2008	National Institutes of Health announce an undiagnosed diseases program to take on unsolved cases of illness that might be solved by unraveling its genetics.

GLOSSARY

anatomy   a branch of morphology that deals with the structure of organisms

antibiotic   a substance produced by or semisynthetic substance derived from a microorganism and able to dilute solution to inhibit or kill another microorganism

antiseptic   opposing sepsis, putrefaction, or decay; preventing or arresting the growth of microorganisms

antitoxin   an antibody that is capable of neutralizing the specific toxin (as a specific causative agent of disease) that stimulated its production in the body and is produced in animals for medical purpose by injection of a toxin or toxoid with the resulting serum being used to counteract the toxin in other individuals

aseptic   preventing infection

bacteriophage   a virus that infects bacteria

chemotherapy   the use of chemical agents in the treatment or control of disease

dormant   marked by a suspension of activity; temporarily devoid of external activity

hemoglobinuria   the presence of free hemoglobin in the urine

microbe   microorganism, germ

nitrous oxide (N_2O) a colorless gas that when inhaled produces loss of sensibility to pain preceded by exhilaration and sometimes laughter, that is used especially as an anesthetic in dentistry and as a fuel, and that is an atmospheric pollutant and greenhouse gas produced by combustion; known as laughing gas

nucleotide   any of several compounds that consist of a ribose or deoxyribose sugar joined to a purine or pyrimidine base and to a phosphate group and that are the basic structural units of nucleic acids

pasteurization   partial sterilization of a substance, especially a liquid, at a temperature and for a period of exposure that destroys objectionable organisms without major chemical alteration of the substance

physiology   a branch of biology that deals with the functions and activities of life or of living matter and of the physical and chemical phenomena involved

pitchblende   a brown to black material that consists of massive uraninite, has a distinctive luster, contains radium, and is the chief ore-mineral source of uranium

poliomyelitis   an acute infectious disease caused by poliovirus and characterized by fever, motor paralysis, and atrophy of skeletal muscles often with permanent disability and deformity and marked by inflammation of nerve cells in the anterior gray matter in each lateral half of the spinal cord

postulate   a hypothesis advanced as an essential presupposition, condition, or premise of a train of reasoning

prion   a protein particle that lacks nucleic acid and has been implicated as a source of various neurodegenerative diseases (i.e., bovine spongiform encephalopathy)

proprietary   possessing, owning, or holding exclusive rights to something

septicemia   invasion of the bloodstream by virulent microorganisms and especially bacteria; blood poisoning

serum   a watery portion of an animal fluid remaining after coagulation

spirochete   any of an order of slender spirally undulating bacteria

suffragette   a woman who advocates voting rights for women

synthetic   something resulting from synthesis rather than occurring naturally

tomography   a method of producing a three-dimensional image of the internal structures of a solid object by the observation and recording of the differences in the effects on the passage of waves of energy impinging on those structures

vector   an organism that transmits a pathogen

viroid   any of two families (*Pospiviroidae* and *Avsunviroidae*) of subviral particles that consist of a small single-stranded RNA arranged in a closed loop without a protein shell and that replicate in their host plants where they may or may not be pathogenic

virus   the causative agent of an infectious disease; also any of a large group of submicroscopic infective agents that are regarded as either as extremely simple microorganisms or as extremely complex molecules, that typically contain a protein coat surrounding an RNA or DNA core of genetic material but no semipermeable membrane, that are capable of growth and multiplication only in living cells, and that cause various important diseases in humans, lower animals, or plants

X-ray   any of the electromagnetic radiations that have an extremely short wavelength of less than 100 angstroms and have the properties of penetrating various thicknesses of all solids, of producing secondary radiations by impinging on material bodies, and of acting on photographic films and plates as light does

ABOUT SCIENCE AND HISTORY

Diamond, Jared. *Guns, Germs, and Steel: The Fates of Human Societies.* New York: W. W. Norton, 1999. Diamond places the development of human society in context, which is vital to understanding the development of medicine.

Hazen, Robert M., and James Trefil. *Science Matters: Achieving Scientific Literacy.* New York: Doubleday, 1991. A clear and readable overview of scientific principles and how they apply in today's world, including the world of medicine.

Internet History of Science Sourcebook. Available online. URL: http://www.fordham.edu/halsall/science/sciencsbook.html. Accessed July 9, 2008. A rich resource of links related to every era of science history, broken down by disciplines, and exploring philosophical and ethical issues relevant to science and science history.

Lindberg, David C. *The Beginnings of Western Science,* 2nd ed. Chicago: University of Chicago Press, 2007. A helpful explanation of the beginning of science and scientific thought. Though the emphasis is on science in general, there is a chapter on Greek and Roman medicine as well as medicine in medieval times.

Roberts, J. M. *A Short History of the World.* Oxford: Oxford University Press, 1993. This helps place medical developments in context with world events.

Silver, Brian L. *The Ascent of Science.* New York: Oxford University Press, 1998. A sweeping overview of the history of science from the Renaissance to the present.

Spangenburg, Ray, and Diane Kit Moser. *Science Frontiers: 1946 to the Present,* rev. ed. New York: Facts On File, 2004. A highly readable book with key chapters on some of the most significant developments in medicine.

ABOUT THE HISTORY OF MEDICINE

Ackerknecht, Erwin H., M.D. *A Short History of Medicine,* rev. ed. Baltimore, Md.: Johns Hopkins University, 1968. While there have been many new discoveries since Ackerknecht last updated this book, his contributions are still important as they help the modern researcher better understand when certain discoveries were made and how viewpoints have changed over time.

American Red Cross. The Web site of the American Red Cross has a very good history of Clara Barton. Available online. URL: http://www.redcross.org. Accessed February 10, 2009.

Andermann, Anne Adina Judith. "Physicians, Fads, and Pharmaceuticals: a History of Aspirin." *McGill Journal of Medicine* vol. 2 (1996): pp. 115–120. This journal article provides a fascinating background on how aspirin developed and was used.

Buchan, William. *Domestic Medicine,* 2nd ed. London: Royal Society, 1785. Available online. URL: http://www.americanrevolution.org/medicine.html. Accessed January 10, 2009. This book provides a contemporary account of the medical beliefs of the late 1700s.

Clendening, Logan, ed. *Source Book of Medical History.* New York: Dover Publications, 1942. Clendening has collected excerpts from medical writings from as early as the time of the Egyptian papyri, making this a very valuable reference work.

Cochrane, A. L. *One Man's Medicine: The Autobiography of Archie Cochrane.* London: Wiley Blackwell (Memoir Club), 1989. This is Cochrane's story of how and why he came up with the idea of evidence-based medicine.

Dary, David. *Frontier Medicine: From the Atlantic to the Pacific 1492–1941.* New York: Alfred A. Knopf, 2008. This is a new book that has been very well reviewed. Dary outlines the medical practices in the United States from 1492 forward.

Davies, Gill, ed. *Timetables of Medicine.* New York: Black Dog & Leventhal, 2000. An easy-to-assess chart/time line of medicine with overviews of each period and sidebars on key people and developments in medicine.

Dittrick Medical History Center at Case Western Reserve. Available online. URL: http://www.cwru.edu/artsci/dittrick/site2/ Accessed January 10, 2009. This site provides helpful links to medical museum Web sites.

Duffin, Jacalyn. *History of Medicine.* Toronto: University of Toronto Press, 1999. Though the book is written by only one author, each chapter focuses on the history of a single aspect of medicine, such as surgery or pharmacology. It is a helpful reference book.

Gotcher, J. Michael. "Assisting the Handicapped: The Pioneering Efforts of Frank and Lillian Gilbreth." *Journal of Management* 18, 5 (1992). Available online. URL: http://jom.sagepub.com/cgi/content/abstract/18/1/5. Accessed January 21, 2009. The Gilbreths have generally been ignored for their work on behalf of the handicapped; this article does a great deal to remedy that by explaining their very considerable contributions.

Kennedy, Michael T., M.D., FACS. *A Brief History of Disease, Science, and Medicine.* Mission Viejo, Calif.: Asklepiad Press, 2004. Michael Kennedy was a vascular surgeon and now teaches first- and second-year medical students an introduction to clinical medicine at the University of Southern California. The book started as a series of his lectures, but he has woven the material together to offer a cohesive overview of medicine.

Loudon, Irvine, ed. *Western Medicine: An Illustrated History.* Oxford: Oxford University Press, 1997. A variety of experts contribute chapters to this book that covers medicine from Hippocrates through the 20th century.

Magner, Lois N. *A History of Medicine.* Boca Raton, Fla.: Taylor & Francis Group, 2005. An excellent overview of the world of medicine from paleopathology to microbiology.

Medical Discoveries. This Web site provides an alphabetical resource with biographies and other information about important medical milestones. Available online. URL: http://www.discoveriesinmedicine.com. Accessed February 26, 2009.

National Human Genome Research Institute. This is a National Institutes of Health government-sponsored site to provide the public with information about all aspects of research concerning the human genome. Available online. URL: http://www.genome.gov. Accessed February 1, 2009.

Porter, Roy. *The Greatest Benefit to Mankind: A Medical History of Humanity.* New York: W. W. Norton, 1997. Over his lifetime, Porter wrote a great amount about the history of medicine, and this book is a valuable and readable detailed description of the history of medicine.

————, ed. *The Cambridge Illustrated History of Medicine.* Cambridge, Mass.: Cambridge University Press, 2001. In essays written by experts in the field, this illustrated history traces the evolution of medicine from the contributions made by early Greek physicians through the Renaissance, scientific revolution, and 19th and 20th centuries up to current advances. Sidebars cover parallel social or political events and certain diseases.

Red Gold: The Epic Story of Blood. This is a PBS Web site to accompany a series with this title, and the Web site is one of the most extensive and interesting about how and when scientists learned about blood. Available online. URL: http://www.pbs.org/wnet/redgold/history/index.html. Accessed February 20, 2009.

Rosen, George. *A History of Public Health, Expanded Edition.* Baltimore: Johns Hopkins University Press, 1993. While serious public health programs did not get underway until the 19th century, Rosen begins with some of the successes and failures of much earlier times.

Sherrow, Victoria. *Jonas Salk: Beyond the Microscope, Revised Edition.* New York: Chelsea Publishers, 2008. This book is an excellent reference for reading about polio, about Salk, and about how he and Sabin interfaced in their efforts to combat polio.

Simmons, John Galbraith. *Doctors & Discoveries.* Boston: Houghton Mifflin Company, 2002. This book focuses on the personalities behind the discoveries and adds a human dimension to the history of medicine.

Starr, Paul. *The Social Transformation of American Medicine.* New York: Perseus, 1982. The book puts in perspective the changes in the American medical system and how they came about.

Toledo-Pereyra, Luis H. *A History of American Medicine from the Colonial Period to the Early Twentieth Century.* Lewiston, N.Y.: Edwin Mellen Press, 2006. This is an academic book that provides very valuable information about medicine in the 19th century.

United States National Library, National Institutes of Health. Available online. URL: Available online. URL: http://www.nlm.nih.gov/hmd/. Accessed July 10, 2008. A reliable resource for online information pertaining to the history of medicine.

OTHER RESOURCES

Collins, Gail. *America's Women: 400 Years of Dolls, Drudges, Help-mates, and Heroines.* New York: William Morrow, 2003. Collins's book contains some very interesting stories about women and their roles in health care during the early days of America.